JOURNEY ONTO LAND

Life on Land in the Early Permian.

Journey onto Land

COLEMAN J. GOIN

OLIVE B. GOIN

Illustrated by
MARGARET MATTHEW COLBERT

MACMILLAN PUBLISHING CO, INC.
NEW YORK

COLLIER MACMILLAN PUBLISHERS
LONDON

Macmillan Publishing Co., Inc.
866 Third Avenue, New York, New York 10022

Collier-Macmillan Canada, Ltd.

Library of Congress Cataloging in Publication Data

Goin, Coleman Jett (date)
  Journey onto land.

  Includes bibliographies.
  1. Adaptation (Biology) 2. Evolution. I. Goin,
Olive Bown, joint author. II. Title.
QH546.G64   575   73-8580
ISBN 0-02-344210-7

Printing: 1 2 3 4 5 6 7 8   Year: 4 5 6 7 8 9 0

PREFACE

THIS LITTLE BOOK looks at the three groups of organisms that became really terrestrial—the seed plants, arthropods, and vertebrates. Life on land presents quite different problems from those posed by an aquatic existence. These three very diverse groups have had basically the same problems to solve. How they did so is a success story attested by their predominance today.

We are not concerned here with the details of the processes involved in speciation. Discussions of mutation, recombination, isolation, and natural selection are readily available and are familiar to most biology students. Nor do we pay much attention to phylogenies. The data are all too scanty for every group except the vertebrates and are more properly the domain of the paleontologists. We do present in the last chapter speculation on a basic evolutionary mechanism that allowed these three groups to make such a major adaptive shift.

As always, we are indebted to a number of friends for help. We would mention particularly Dr. E. H. Colbert, who read the chapter on the vertebrates, and Dr. Walter B. McDougall, who read the chapter on plants. We wish to thank Deloris Douglas for her careful typing of the manuscript. Finally, Margaret Matthew Colbert, who made the original drawings particularly for this volume, has us very much in her debt.

C. J. G.

O. B. G.

vii

CONTENTS

JOURNEY ONTO LAND

CHAPTER 1 The Rigors of Life on Land

THE JOURNEY ONTO LAND was not an easy one. Life arose in the water and few indeed were the kinds of organisms that were completely successful in making the major shift to life on dry land.

Take a walk through the fields and woods some sunny spring morning. Buttercups gleam among the meadow grasses, bees search for nectar in the clover, you may brush against the web of an orb-weaving spider strung between two bushes. A garter snake may glide quickly across your path, a squirrel scold head down from the trunk of an oak tree, while the loud ringing song of a robin sounds from the upper branches. The dry land is teeming with life. Yet all the living things you have seen during your walk probably are members of only three of the more than thirty major groups of animals and plants that exist in the world today. They are either **arthropods, vertebrates,** or **vascular plants.** It is probable that only a single kind of vertebrate, two kinds of arthropods, one kind of green alga gave rise to all the animals and plants you have encountered.

The descendants of these few pioneers diversified to give rise to the myriads of different species living on dry land today.

AQUATIC VERSUS TERRESTRIAL ENVIRONMENTS

The living things that successfully invaded the land, plant, arthropod, or vertebrate, faced essentially the same problems and the adaptations by which they met these problems show many similarities. To understand these adaptations, we should first look at the problems in response to which they evolved. Just how does a terrestrial environment differ from an aquatic one and what problems do the differences pose?

Water Balance. Water is not only an essential ingredient of all living things, it is their major component. The protoplasm of actively living cells is 60 to 90 per cent water. To live and grow, all organisms must have a readily available source of water. In an aquatic environment, the organism is surrounded by water. It is true that such organisms are often not in osmotic equilibrium with the environment and have evolved special adaptations to maintain the proper amount of water within the body. The number of water molecules per unit volume in an aqueous solution is determined by the number of molecules of other substances dissolved in the water—the greater the number of other molecules, the fewer the water molecules. A living cell in an aquatic environment represents an aqueous solution separated by a semipermeable membrane, the cell membrane, from another aqueous solution, the surrounding medium. The water molecules are in constant motion and can pass through the cell membrane in either direction. If there are more water molecules per unit volume inside the cell than there are outside the cell, then more water molecules move out of the cell than move into it, and if the process continues unchecked, the cell eventually shrivels and dies. Conversely, if there are more water molecules per unit volume outside the cell than there are inside, water moves into the cell; the cell swells and may eventually burst. The movement of water molecules through a semipermeable membrane separating aqueous solutions having different concentrations of dissolved substances is called **osmosis.**

The concentration of dissolved substances in fresh water is lower than the concentration of substances in the bodies of the organisms that live in fresh water. Water from the environment tends to move into their bodies. If they are not to be literally swamped, they must have mechanisms for coping with this influx. Plant cells are surrounded by rigid, nonliving cell walls. As the cell takes in water by osmosis, it swells and exerts pressure on the cell wall (turgor pressure). Since the wall is rigid, it exerts a back pressure on the cell that prevents it from bursting. The increased pressure within the cell increases the outward flow of water molecules until the outflow balances the inward flow and the cell establishes an equilibrium with the environment. Most animal cells lack a rigid outer covering. Animals that live in fresh water, such as aquatic amphibians and freshwater fishes, eliminate the excess water that enters their bodies by excreting copious amounts of very dilute urine.

The concentration of dissolved substances in the sea is about the same as that in the body fluids of many of the organisms that live in the sea, so problems of osmotic regulation do not arise. On the other hand, more dissolved substances are present in sea water than in the bodies of some marine animals. This means that these animals are subject to desiccation as water passes out of their bodies by osmosis. Marine fishes replace the water lost by drinking sea water. They produce very little urine, thereby conserving water, and they eliminate the excess salt taken in by excreting salt through their gills.

The problems of water balance faced by organisms on dry land are much more severe. There are many fewer water molecules present in the air than in the bodies of the plants and animals that live exposed to the air, and desiccation can be very rapid. A jellyfish stranded on the beach soon disintegrates into an amorphous blob. Even animals that are adapted to living in moist soil rather than directly in the water are very susceptible to drying, as the bodies of earthworms stranded on the sidewalk after a night of warm rain attest. They came out of their burrows during the hours of high humidity and were not able to get back in time when the humidity dropped with the coming of day.

The bodies of terrestrial organisms must be protected by a nearly, if not quite, impermeable covering to minimize the water loss to the environment. Yet not all body surfaces can be so protected. Passageways must be left open for the exchange of **oxygen** and **carbon dioxide** with the environment. These gases pass more readily through a moist membrane than through a dry one. Though the surfaces through which gas exchange takes place, such as the inner lining of the lungs of animals or the inside membranes of the leaves of plants, are more or less protected by being inclosed within the body, some water is always lost from them. You can see the water vapor in your breath condense every time you breathe out on a frosty morning. Plants lose water by evaporation from the surface of the leaves. Elimination of fecal materials and excretion of **nitrogenous waste products** are major factors in water loss for many animals. To maintain the amount of water in the body at an acceptable level and to allow for growth, terrestrial organisms must be able to extract water from environments in which water is never so abundant as it is in the lakes, streams, and oceans.

Temperature and Heat. The words "temperature" and "heat" are often used interchangeably in the vernacular, but they are not the same thing. Temperature is a measurement of the average speed of molecules—the faster the molecules in a given body are moving, the higher the temperature of the body. Heat, on the other hand, is a form of energy. It is usually measured in calories. The amount of heat present is determined not only by the average speed of the molecules but also by their number and mass. The temperature in a room is determined by the average speed of the air molecules in the room. If we divide the room in two by putting a partition down the center, the temperature in the two new rooms remains the same as long as the average speed of the molecules is unchanged. But the amount of heat in each of the new rooms is only half that in the original room because each has only half the number of molecules.

Water differs markedly from air in its thermal characteristics. For one thing, it has a much higher specific heat. This means that it takes more heat energy to raise the temperature of water than it does to raise the temperature of air. The specific heat of water is 1; that is, it takes 1 calorie of heat to raise 1 gram of water 1 degree Celsius (formerly called centigrade). The specific heat of air is about 0.28. One calorie of heat will raise the temperature of 1 gram of air nearly 4 degrees Celsius. Evaporation takes place more or less continually from the surface of a body of water, and water has the highest known latent heat of evaporation: 536 calories per gram are used during evaporation. Water thus gains and releases heat slowly, and it is much less subject to rapid and extreme fluctuations in temperature than air is. Large bodies of water moderate the air temperature over nearby land. It is cooler on the shore of a lake on a hot summer afternoon than it is farther inland, and it is also warmer during cold weather. For this reason citrus growers in Florida prefer to plant their groves on the shores of lakes because there is less danger of frost damage to the fruit and trees during the cold waves that move down the peninsula in winter. As the pioneer terrestrial organisms moved away from the water, they were obliged to develop mechanisms to protect themselves from extremes of heat and cold unknown to their aquatic ancestors.

Support. Per unit volume, water weighs about 773 times as much as air. This means that it offers much more support for the bodies of plants and animals. If a terrestrial plant is not to lie recumbent on the ground, if a terrestrial animal is to do more than wriggle along the surface, each must have tissues that provide firm support for the body.

Nutrient Supply. The main groups of terrestrial organisms—seed plants, arthropods, and vertebrates—all gave rise to some lines that returned to the water and became secondarily aquatic. These **secondary aquatics** include such forms as water lilies, duckweed, pickerel weed, diving beetles, water spiders, sea snakes, and whales. In contrast to these are the **primary aquatics,** animals and plants whose ancestors were never terrestrial. Primary aquatic plants may be attached to the substrate by a holdfast, they may float and drift with the currents, or they may be equipped with flagella and be capable of independent movement (some small single-celled or colonial forms, for example). In any case, they are surrounded by the substances they need for life, growth, and reproduction. Not only oxygen and carbon dioxide but other requisite inorganic substances, such as nitrates, phosphates, and salts of calcium and potassium, are dissolved in the water. They reach the plants in sufficient supply by the movement of water currents or through the active or passive movements of the plants themselves.

On land the picture is different. Oxygen and carbon dioxide

are present in the air, but the most dependable supplies of water and of inorganic nutrients such as nitrates and phosphates lie below the surface of the ground. Terrestrial plants need mechanisms (root systems) by which they can tap these sources of supply. By their roots they are anchored to the ground; they are incapable of independent locomotion, and they do not float on air currents as seaweeds float on water.

The primary aquatic animals, like the plants, may be either anchored to the substrate for at least part of their life span, or float passively, or be able to move about under their own power, either by crawling along the bottom or by swimming. Like the plants, they make use of the oxygen dissolved in the water around them.

But unlike plants, animals are not capable of manufacturing their own food from inorganic substances. They must eat either plants or other animals. Attached animals such as sea anemones, sea lilies, barnacles, and oysters depend on currents of water to bring food particles to them. A terrestrial animal attached to the ground would soon exhaust the food supply in its immediate vicinity. Winds over land are less dependable in their flow than are water currents, and because air is less dense than water, it is less capable of transporting animals and plants. To get their food, all animals that live on land must be able to move about. Lacking roots, they must usually depend on other sources of water than the ground water available to the plants. Furthermore, since the animals and plants on which they feed are protected by tough, waterproof outer layers, they must have some means, such as teeth, horny jaws, or piercing and sucking mouth parts, of breaking through these coverings.

Sense Organs. Animals that are capable of active locomotion need to be well equipped with sense organs. They have to be able to see where they are going, to avoid obstacles; to locate food, water, and potential mates; and to recognize the approach of predators. The sense organs of active aquatic animals are designed to function in water. In fishes, the eyes are constantly washed by water; the chemical sense organs respond to substances dissolved in water and may be located on the outer surfaces of the body as well as in the mouth and nasal cavities; currents and vibrations in the water are perceived through a series of lateral line sense organs; the sense of hearing is absent or at best poorly developed. All these sensory systems were modified in terrestrial vertebrates to adapt them for functioning in air rather than water. The cornea, the outer coat of the eye, becomes opaque when it dries out. Tear glands and eyelids evolved to keep the cornea moist and washed free of dust particles. The exposed body surfaces of terrestrial vertebrates are covered by a layer of dry skin to conserve water; the chemical sense organs are restricted to the mouth and nasal cavities. The lateral-line sense organs are no longer needed, but hearing becomes more important and the ear must be adapted to receive vibrations in the air.

Internal Transport. Many primarily aquatic organisms are either very small or are flattened in shape so that no cell in the body is very far from the surrounding water. The needs of the cells to take in nutrients and to rid themselves of waste products can be met by diffusion from one cell to another and between the surface cells and the medium. The complexity of organization needed to meet the complex demands of the terrestrial environment required in both plants and animals a diversification of tissue types. Organisms composed of many different tissues are in general bulkier than those composed of only a few types. Many of the individual cells are too far away from the external environment for efficient exchange by simple diffusion. Furthermore, since the bodies of terrestrial organisms are largely protected by more or less impermeable surfaces, exchange with the environment takes place only in localized areas, not over the whole body surface. Both animals and plants that live on land need efficient systems of **internal transport.** The ancestors of terrestrial animals had such systems, but the plants apparently evolved them after they came ashore.

Waste Disposal. All animals produce nitrogenous wastes in the course of their metabolic activities. Usually these wastes are in the form of **ammonia** (NH_3). Ammonia is highly soluble in water and it is also very toxic to living tissues. Small aquatic animals are able to rid their cells of ammonia by diffusion; the surrounding water quickly dissolves it and carries it away from their body surfaces. Diffusion is not sufficient to meet the needs of larger aquatic animals. Fresh-water fishes and larval and aquatic amphibians use the abundant supply of water that enters their bodies by osmosis to dissolve and carry off the ammonia. They produce large quantities of very dilute urine. Marine bony fishes, which must conserve body water, excrete most of the ammonia through their gills and again the surrounding water carries it away from their body surfaces. None of these mechanisms are available to animals that live on land. They cannot afford to expend large amounts of water as the freshwater fishes do, and there is no water in the surrounding medium to dissolve and carry off the toxic ammonia if it were excreted through special body surfaces. Terrestrial animals have had to evolve other means of coping with the problem of nitrogenous wastes. (Plants apparently do not face this problem. They are able to recycle simple nitrogenous compounds and they take in from the soil only as much of these compounds as they need for new growth.)

Reproduction. Most animals and plants reproduce sexually. They form special sex cells, the **gametes.** Two of these cells fuse to form a single cell, the **zygote** or **fertilized egg.** Usually the two cells that unite to form the zygote are different. One, the **ovum,** is larger and is not capable of independent movement. The other, the **sperm,** is smaller and is mobile. Frequently, the ovum and sperm are produced by different individuals. Even when they are produced by the same individual, they are formed in different parts of the organism. The sperm must be able to travel to the ovum to fertilize it. Sperm of primary aquatic organisms are equipped with one or more flagella and can actively swim through the water to reach the ovum. On land, not only must both ovum and sperm be protected from desiccation, but the sperm must have some way to reach the ovum. If it is to swim, as the sperm of terrestrial animals do, it must be provided with a fluid medium; if it cannot reach the egg by swimming, as in the higher plants, some other mechanism must have evolved to bring it into contact with the ovum.

The zygote of a primary aquatic plant can be released directly into the water. It is in no danger of desiccation. Furthermore, the nutrients it needs for growth and development are present in the surrounding medium. It is capable of an independent existence right from the start. This is not possible for the zygote of a terrestrial plant. It must be protected from desiccation until it has developed enough to be able to procure its own water, and it must be provided with a food supply that will allow it to grow and differentiate and so form the tissues it needs to take in nutrients and manufacture its own food. It must live for awhile as an **embryo,** that is, as a developing organism that receives its nourishment either directly from its mother's body or from a stored food supply.

The zygote of an aquatic animal is no more capable than the

adult of using small, inorganic molecules in the water to manufacture its own food. Until it can fend for itself, it too must pass through an embryonic stage. Its food is usually provided by **yolk** stored in the ovum before fertilization. The animals that moved onto land did not need to modify their life histories to include a period of embryonic development. But they did need to evolve mechanisms to protect their delicate embryos from desiccation.

ADVANTAGES OF TERRESTRIAL ENVIRONMENT

With all the hazards that living on dry land presented to organisms that, at least at first, were ill equipped to meet them, it may seem strange that terrestrial life evolved at all. Why should any species leave the water, to which it was well adapted, and move out onto the strange, barren, inhospitable land? What compensating advantages could the new life offer to make the venture the success that it was? Some temporary advantages there always are for the first pioneers into any essentially uninhabited area. There are no competitors, no closely related species making similar demands on the environment for the necessities of existence, for food and water and living space. There are no enemies. For the first plants, there were no plant-hungry herbivores to feed on them; for the first animals, no meat-hungry carnivores. But these advantages were only temporary. For both plants and animals, the invasion of the land was followed by a rapid diversification of species. Competition grew keen and enemies soon appeared. There was, though, one great and continuing advantage offered by the terrestrial environment—an abundant and remarkably constant supply of **respiratory gases.**

Green plants are **photosynthetic.** By means of **chlorophyll,** the substance that gives them their green color, they are able to use the energy of light to split water molecules. They combine the hydrogen thus released with carbon dioxide to form sugar. In this process, both oxygen and water are produced as by-products:

$$6CO_2 + 12H_2O + \text{energy} \rightarrow C_6H_{12}O_6 + 6O_2 + 6H_2O$$

By this means, energy is stored in the sugar molecule. It is this energy that the plant uses to carry on its many other activities, to build its proteins and nucleic acids, to maintain itself and to grow. But before the energy can be made available, the sugar molecule must be broken down again by the process of **cellular respiration:**

$$C_6H_{12}O_6 + 6O_2 \rightarrow 6CO_2 + 6H_2O + \text{energy}$$

Here oxygen is used and the by-products are carbon dioxide and water.

Photosynthesis goes on only during the daylight hours, but cellular respiration continues both day and night. During the day the plant must take in carbon dioxide, but at the same time it produces more than enough oxygen for its own needs; the excess is given off as a waste product. At night the plant must take in oxygen from the environment. In total, though, plants produce much more oxygen than they consume. Animals, like plants, are constantly carrying on cellular respiration; the primary source of the oxygen they use is probably the excess produced by the green plants.

Oxygen is only slightly soluble in water. Almost all the oxygen produced by aquatic plants passes into the atmosphere. On the other hand, oxygen is a major component of the air (about 21 per cent by volume). It is over twenty times as abundant in the air as it is in lakes, rivers, and oceans. Furthermore, the amount of oxygen in the air at any given altitude is remarkably constant over the whole globe, whereas the amount of oxygen in water is subject to great fluctuations. Winds, waves, and currents mix air with the water; cold water holds more dissolved oxygen than warm water does; and the respiratory activities of aquatic plants and animals and the decay of organic materials remove oxygen from the water.

In contrast to oxygen, carbon dioxide is very soluble in water and it is constantly being added to water from soil and underground sources by the respiratory activities of both plants and animals, and by the processes of organic decay. The amount of carbon dioxide in the air is very low (about 0.03 per cent by volume) and it may be a limiting factor, determining how much plant growth can take place. But even though it is relatively more available in water, still, as with oxygen, the amount of dissolved carbon dioxide is subject to wide fluctuations. Aquatic plants cannot depend on a steady, constant supply.

An abundance of oxygen and a constancy in the available amounts of both oxygen and carbon dioxide were the rewards awaiting those organisms that succeeded in establishing a foothold on land.

PREADAPTATION AND POSTADAPTATION

We have talked about the many adaptations that organisms have evolved in becoming successful inhabitants of dry land. This does not mean the animals and plants first moved out on land and then developed all the characteristics they needed to live on land. While they were still aquatic, they must have had some traits that allowed them to survive, at least temporarily, away from the water. Suppose, for example, that the aquatic ancestors of the first land plants lived in a place where they were subject to fluctuations in the water level. Of the plants stranded on shore when the water receded, those best able to withstand temporary drying would be the ones most apt to survive and pass their traits on to their descendants. Through millenia of time the combined forces of mutation, recombination, and natural selection would act to improve the ability to resist desiccation, but some ability had to be there to begin with. The plants were **preadapted** to withstand temporary drying. But they had not evolved this ability in anticipation of becoming terrestrial. The morphological and/or physiological characters (we do not know for sure what they were) that enabled the ancestors of the terrestrial plants to survive being stranded on the shore must have been useful to them in an aquatic environment.

A **preadaptation** is a character, presumably evolved in response to the requirements of one environment, that enables the organism to invade a different environment. It may serve a different function in the second environment from the one it had in the first. Arthropods are covered with a tough, jointed, outer skeleton which protects the terrestrial insects and spiders from desiccation, but it is doubtful if this was its original function. Living aquatic arthropods and the extinct aquatic forms from which the terrestrial ones probably evolved have such exoskeletons, too. Muscles are attached to the plates that form the skeleton, and the combined musculoskeletal system endows the arthropod with a locomotor apparatus that permits more rapid and diversified movements than are possible for most other aquatic invertebrates. (The agility of the arthropods is, of course, just as useful to them on land.) It seems probable, then, that the hard, impermeable plates that make up the arthropod external skeleton evolved as places for muscle attachment. That they could also protect the animal from desiccation was coincidental. But their presence preadapted the arthropods for the journey onto land.

All the organisms that made this journey must have had some preadaptations to start with. But this does not mean that they evolved all the adaptations they needed for success on land in the water before they started out. Once an animal or plant species establishes a foothold in a radically new environment, it is exposed to very strong selection pressure. Those of its descendants that evolve new and better adaptations (**postadaptations**) survive and multiply. Those that do not are eliminated. Preadaptations allow the organism to enter the new environment; postadaptations determine its success there.

SELECTED REFERENCES

DELAVORYAS, THEODORE. *Plant Diversification.* New York, Holt, Rinehart and Winston, Inc., 1966.

FINGERMAN, MILTON. *Animal Diversity.* New York, Holt, Rinehart and Winston, Inc., 1969.

MCALESTER, A. L. *The History of Life.* Englewood Cliffs, New Jersey, Prentice-Hall, Inc., 1968.

MULLER, W. H. *Botany: A Functional Approach,* 2nd ed. New York, Macmillan Publishing Co., Inc., 1969.

ODUM, E. P. *Fundamentals of Ecology,* 3rd ed. Philadelphia, W. B. Saunders Company, 1971.

CHAPTER 2

The Seed Plants

PLANTS WERE THE FIRST organisms to come ashore. They had to be. Only they are able to make use of inorganic materials to manufacture food. Animals could not move far from the food resources of the water and so really establish themselves on land until a terrestrial food supply had been made available to them by the plants. It is no surprise, then, that the earliest fossils found in terrestrial sediments are those of plants. These fossils appear in deposits laid down in the Late Silurian period, more than 400 million years ago (see the Appendix for the geologic time scale). What were these early land plants like and where did they come from?

The **plant kingdom** is divided into a number of major groups called **divisions.** Most of these divisions comprise plants that are collectively called **algae.** All the primary aquatic plants living today, the pond scums of fresh water, the sea weeds of the oceans, belong to one or another of the divisions of algae. An alga may consist of only a single cell, or it may be made up of many cells. The cells may form long, sometimes branching, filaments or broad, leaflike expansions called blades. There may be some cellular differentiation, but even the largest algae have fewer kinds of cells than are found in terrestrial plants. The reproductive cells are not sheltered in special or-

17

gans formed of nonreproductive cells and there is no embry-onic stage.

Algae show a variety of patterns of sexual reproduction. Remember that in ordinary cell division (**mitosis**) each of the **chromosomes** duplicates itself and each of the daughter cells receives a set containing one copy of each chromosome. Among animals that reproduce sexually, gametes are formed by a special kind of cell division (**meiosis**) in which the chromosomes form pairs and each daughter cell receives a set containing one member of each pair. As a result, the gamete is **haploid;** that is, it has only half the number of chromosomes per nucleus that the parent organism has. The original (**diploid**) number is restored when two gametes fuse to form a fertilized egg or zygote. The diploid zygote then grows to form the adult; only the gamete stage is haploid.

Sexual reproduction in algae is not so simple. Some of them reproduce the same way the animals do: the diploid zygote grows into the mature plant, which then produces gametes by meiosis. In other algae the diploid zygote divides by meiosis to give rise to a set of haploid cells called **spores.** Each spore is capable of growing into a mature plant which is haploid and which produces haploid gametes by mitosis. These fuse to form the zygote, which is the only diploid stage in the life cycle. Still other algae show a combination of these two types of reproduction. The zygote grows into a mature diploid plant which produces haploid spores by meiosis. The spore grows into a mature haploid plant which produces haploid gametes by mitosis. This succession—haploid plant–diploid plant–haploid plant—is called **alternation of generations** (see Fig. 2-1).

You may find it hard to keep in mind the distinction between a spore and a gamete. Most biology students are first introduced to the mechanics of gamete formation with a de-

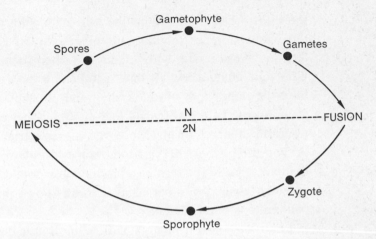

Fig. 2-1. Sexual reproduction with alternation of generations. (From C. J. Goin and O. B. Goin, *Man and the Natural World,* Macmillan Publishing Co., Inc., 1970. © Copyright by Coleman J. Goin and Olive B. Goin, 1970.)

scription of what happens in animals and are accustomed to thinking of gametes as being the direct products of meiosis. Think instead of a gamete as a haploid cell that fuses with another to form a diploid zygote. Whether it is a direct product of meiosis or of mitosis depends on whether the parent plant was diploid or haploid. A reproductive spore, on the other hand, while a direct product of meiosis, never fuses with another spore to produce a zygote; it grows directly into a mature haploid organism.

In plants that have alternation of generations, the diploid and haploid stages are given different names. The diploid plant, which produces spores by meiosis, is called the **sporophyte.** The haploid plant, which produces gametes by mitosis, is called the **gametophyte.**

Botanists believe that the land plants evolved from some ancient member of the division Chlorophyta, the green algae, of which some 6500 species are living today. The pond scums, the slimy green floating mats that appear in still waters during the warm summer months, are masses of green algae. Sea lettuce, whose broad, leafy blades are often seen in the inter-tidal zone along a rocky shore, also belongs here. These algae have the same kinds of chlorophyll and other pigments as are found in the land plants, and like the latter they store food in the form of starch or oil. Moreover, all terrestrial plants show in their life histories at least some traces of alternation of gen-erations, and this pattern of reproduction is found in a number of green algae. No other algal division shows this particular combination of characters.

Let us turn now from an aquatic plant that had the potential for evolving into a terrestrial one to a plant that is obviously well adapted to living on land—a pine tree (see Fig. 2-2). You

Fig. 2-2. A pine tree, *Pinus ponderosa.*

can see at a glance that a pine tree is a large plant well dif-ferentiated structurally. It has an extensive root system, a tall, woody, erect stem, branches, needles, and cones. It can thrive in dry soil well away from any open body of water. Despite the old song, it is seldom lonesome; it usually grows in the company of many other trees of the same kind; indeed, a forest of pine trees often extends for hundreds of miles.

A closer examination of the structure of a pine tree shows why it is so successful on dry land. The outside of the trunk is covered with a layer of **cork,** a fairly thick, hard, nonliving tissue, formed of what were once living cells. Now only the cell walls remain, thickened by deposits of suberin, a waxy, waterproof material. This outer layer of cork forms a tough, impervious covering that protects the living tissues within, both from the loss of essential water to the atmosphere and from attack by voracious hordes of plant-hungry animals. Few insects and even fewer vertebrates are equipped to penetrate the pine tree's protective coat.

Under the cork is a layer of **phloem,** a living **conductive tissue** a number of cell layers thick. The individual transporting cells are elongated, with tapering, overlapping ends. They are called **sieve cells** because their walls are perforated by holes through which pass cytoplasmic strands connecting one cell with another. In this way continuous streams of protoplasm are formed which extend from the needles through the twigs and branches, down the trunk, and into the roots of the tree. Food manufactured in the needles moves along these streams to provide nourishment for the cells in other parts of the tree, which, lacking chlorophyll, are not able to make their own.

Pine trees keep their needles throughout the winter, but many trees lose their leaves and perforce stop manufacturing food. Biennial and perennial herbaceous plants (nonwoody plants that live for two or more years) may die down completely above ground. These plants store excess food manufactured during the summer in roots and stems. A sweet potato is a root expanded for storage and an Irish potato is an underground storage stem. Many of our other common vegetables, such as beets, carrots, and turnips, are similar food reservoirs for the plants. When spring comes, this stored food moves upward to supply the energy and materials needed for new growth. The sweet rising sap of a sugar maple is food being carried up to the tips of the twigs to nourish the swelling buds.

Transport in the phloem is thus a two-way process. Food can move up as well as down. Just how this transport system works is one of the great remaining mysteries of botany. Partial pressure of food molecules may have something to do with it. When the sun is shining brightly and the tree is in full leaf, much more sugar is being made than is being used in the chlorophyll-bearing cells. The concentration of sugar molecules in them is higher than in the adjacent phloem cells and the molecules tend to move down the concentration gradient. But the movement of food through the phloem from one part of a plant to another takes place much more rapidly than can be accounted for by the simple diffusion of molecules. A mass flow of the sugar molecules through the protoplasm of the phloem cells and a streaming of the whole protoplasmic content of the cells have both been suggested, but these processes are also too slow. Some kind of rapid transport system must be at work.

Phloem is one of the **vascular** or **conductive tissues** that characterize the land plants. By it many cells of the plant body are freed from the necessity of manufacturing their own food and can specialize for other functions. Thus root cells, buried in darkness underground, can survive and carry on the essential business of procuring water and minerals for the whole tree.

Phloem and cork together form the **bark** of the tree. Pioneers in this country frequently cleared fields for cultivation by girdling, that is, by removing a strip of bark completely around the trunk of each tree. This destroyed the phloem passage between leaves and roots; the roots, deprived of food, died; and the rest of the tree, deprived of water and minerals, did not long survive.

Beneath the bark of the pine lies a layer of tissue called **cambium.** It is made up of small cubical or elongated cells with very thin walls and without the usual large central vacuoles found in most plant cells. Cambium is a **meristematic tissue;** this means that the cells, in contrast to other cells in the plant body, continue to divide mitotically. Bundles of meristem (apical meristem) are also found at the tips of the twigs and at the ends of the roots. Cells produced in these regions by mitotic divisions differentiate to form the other tissues. The cells that are to become sieve cells elongate; develop their thick, pitted, cell walls; and lose the ability to divide. All growth in height of the tree and in the length of the branches and roots takes place at the tips. However, there remains in the trunk a sheet of meristem between the phloem and the inner tissues, the lateral meristem or cambium. The cells here divide laterally. The outer layer gives rise to new phloem cells. As these cells are added to the bark, it increases in thickness. Since the layer of cork is composed of dead, thick-walled cells that can neither grow nor divide nor stretch, the outer bark cracks, splits, and flakes off. New cork is formed by meristematic cells present in the phloem.

Beneath the cambium of the pine lies the second vascular tissue, the **xylem.** The conductive tubes here are formed of elongated, tapering, overlapping cells called **tracheids.** They, too, have thickened, pitted walls, and the chambers within are connected. The cells themselves die at maturity and only the cell walls remain. Through their interconnected chambers, water and minerals ascend from the root tips through the trunk and branches to the needles at the tips of the twigs. This is a one-way passage (water does not flow backward down the trunk), and the transport mechanism is better understood than it is for phloem. Water molecules have a strong attraction for one another. The tracheids extend out into the needles, where water is used by the food-manufacturing cells. Water passes into these cells from the tracheids by osmosis, and as each molecule moves onward it exerts a pull on those behind. And so the whole column is hauled upward, even to the top of the tallest redwood.

As the cambium layer produces new sieve cells on the outside, it also produces new tracheids on the inner side. When warmer days come in the spring and new growth starts, the xylem cells produced by the cambium are large with relatively thin walls. In a cross section of the trunk, they appear as a light ring. Growth slows during the summer, the cells formed are smaller with thicker walls, and they appear darker. In this way, concentric rings of light grading into dark are added to the trunk with each growing season. They form the wood of the pine. Only the outer, younger layers of xylem (the sapwood) transport water; resin accumulates in the older, inner layers (the heartwood), making it darker and harder. Resin ducts extend into other parts of the tree, even into the needles. The part the stored resin plays in the life of the pine is not clear. Probably it discourages attack by insects and by decay bacteria, and it may also help cut down water loss from the needles. Figure 2-3 shows a cross section of a pine trunk.

Bundles of vascular tissue (**veins**) extend out into the needles, which are the leaves of the pine. Water diffuses from the xylem of the vein into surrounding chlorophyll-bearing cells (the **mesophyll**), where food is manufactured, and sugar in turn moves from these cells into the phloem system. The mesophyll of the needle is surrounded by a layer of thick-walled protective cells. It is this layer that makes the pine needle so stiff and tough. The whole needle is covered by a layer of

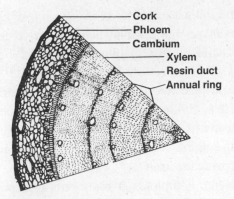

Fig. 2-3. Cross section of a pine stem.

epidermal cells, which secrete a thick, waxy outer cuticle. Sunk into the epidermal layer are a number of openings, the **stomata;** they provide a passageway by which carbon dioxide can reach the mesophyll cells and the oxygen produced by photosynthesis can escape into the atmosphere.

But this arrangement, so necessary for the survival of the pine, also poses a threat. The surfaces of the mesophyll cells must be kept moist to allow for the passage of the gases. Water evaporates from these surfaces and moves out through the stomata (**transpiration**); the route that allows the carbon dioxide needed for photosynthesis to reach the cells also permits the escape of the equally essential water. The pine tree must take in through its roots not just the water needed for the maintenance and growth of its tissues but also enough to compensate for the evaporative water loss from the needles. Pine trees can grow in quite dry regions; they also grow in northern climates and on the higher slopes of mountains, where little water can be absorbed from the frozen ground during the cold winter months. And they keep their needles throughout the year, and with them the danger of desiccation. These needles are really leaves that have been modified by the forces of natural selection to keep water loss at a minimum. The narrow shape, the thick-walled protective cells, the heavy cuticle, the sunken stomata, all serve to retard the transpiration rate.

The mature pine tree is well adapted structurally to life on dry land. How about reproductively? How does fertilization take place away from the water and how is the delicate embryo protected and nourished until it grows roots and needles and is able to provide for itself?

The pine tree is diploid: it represents the sporophyte stage of the life history. Its reproductive organs are the cones, which grow at the tips of the twigs. They are of two kinds, male and female. The female cone develops during the late spring or summer as a central stem surrounded by series of leaflike scales. On the upper surface of each scale are two structures called **ovules.** An ovule has a tough outer covering, the integument, which is pierced by a single tubular opening, the **micropyle.** The integument surrounds a mass of tissue, the **megasporangium,** within which lies a single large cell, the **megaspore mother cell.** ("Megaspore" means "big spore," indicating that the female spore is larger than the male spore.) This cell divides by meiosis; of the four haploid daughter cells, three disintegrate. The single remaining haploid megaspore then divides mitotically a number of times to form a multicellular structure, the **megagametophyte** (meaning "big gamete-forming plant"). This is all that is left of the haploid gametophyte stage that in the algae exists as a separate plant. It remains within the protective ovule of the parent pine and never takes up an independent existence. At the end of the megagametophyte nearest the micropyle, from two to five structures, called **archegonia,** develop. Each contains a single egg.

The male cones are smaller than the female ones. Each scale has two **microsporangia** (meaning "small spore cases") on its lower surface. These contain a number of **microspore mother cells** which divide by meiosis to produce the haploid **microspores.** The pine tree thus has two kinds of spores and is said to be **heterosporous.** Each microspore divides mitotically twice. Two of the daughter cells degenerate; the other two, called the **generative cell** and the **tube cell,** represent a two-celled gametophyte. The gametophyte is surrounded by a protective casing with a wing-shaped projection on each side, the whole forming a **pollen grain.** The microsporangia rupture to release the pollen grains, which are caught up by air currents and, if a strong wind is blowing, may be carried for many miles. By chance some may fall on the scales of a female cone. The advantage to the pine of growing in the company of many others of its kind is obvious—the more pine trees around releasing pollen into the air, the greater the chance that some pollen grains will settle on the female cones. If a pollen grain comes to rest near the micropyle of a megasporangium, it is caught in a drop of sticky fluid exuded by the micropyle and drawn down to the megasporangium as the fluid evaporates. The tube cell elongates, forming a tube that grows toward the megagametophyte. Meanwhile the generative cell has divided to form a **stalk cell** and a **body cell.**

Pollination, that is, the arrival of the pollen grain at the micropyle, takes place when the female cone is young and still developing. Over the summer, as the pollen tube pushes slowly inward, the archegonia form. Then winter comes; growth of the pollen tube practically stops and there is no further development of the megagametophyte. Not till the following spring is there further progress toward fertilization. Then the pollen tube resumes growth toward one of the egg cells; the body cell divides to form two **sperm nuclei;** the tube bursts, releasing the sperm; and one of the sperm nuclei enters and fuses with the nucleus of the egg to form the diploid zygote. Remember that the various nuclei of the pollen grain, like those formed by the divisions of the megaspore, are haploid and are produced by mitosis. The land plants are said to show traces of the alternation of generations found in some of the green algae because, between meiosis and fertilization, these multicellular, haploid structures are formed. But these small, inconspicuous gametophytes, totally dependent on the parent tree, are a far cry from the independent, coequal gametophytes of the primarily aquatic algae.

Once fertilization has taken place, the zygote begins the cell divisions that result in the formation of the embryo. As it grows, it pushes into the tissue of the gametophyte; it gets its nourishment by digesting and absorbing these tissues. The mature embryo has embryonic needles and rudiments that will develop into the roots, stem, and growing shoot of the seedling pine. It is surrounded by the **endosperm,** the food-storage tissues formed by the megagametophyte. The integument of the megasporangium becomes a tough, protective coat. This then is the seed of the pine—an embryo plant surrounded by a food supply and protected by a hard outer covering. Growth of the embryo then stops for a time. The seeds are shed when the ripe pine cone opens. They may fall to the ground under the parent tree or be carried some distance away by the wind. They remain dormant on the soil until the "right" conditions arrive. Then the seed absorbs water and swells, the seed coat bursts, and the embryo develops rapidly into a seedling pine.

The pines and other cone-bearing trees, such as spruce, cypress, and redwood, belong to a group known as the **gymno-sperms,** a word meaning naked seeds. The majority of land plants are **angiosperms** (meaning covered seeds). These are the flowering plants—lilies and apples and daisies and grasses and a host of others. In them the developing gametophytes are enclosed in a complex and often very conspicuous structure, the flower. This not only provides additional protection for the gametophyte and embryo but has allowed the angiosperms to evolve mutual benefit relationships with animals. When the flowers are small and inconspicuous, as in the grasses, oaks, and willows, pollen is usually transferred by the wind as it is in the pine. But plants with showy, usually sweet-smelling, flowers often depend on animal **pollinators.** The flowers usually produce a sugary nectar; the animals (mostly bees but also others, such as butterflies, bats, and hummingbirds) visit the plants to gather food; in so doing they brush against the **stamens,** which are the pollen-producing part of the flower. Some of the pollen clings to the pollinator and is carried to the next flower visited. Here it rubs off on the **stigma,** from which the pollen tube grows down to the ovule. This is a much surer method of achieving pollination than is dependence on wind currents. It is probably one of the main reasons that the angiosperms are even more successful land plants, more widespread and numerous, than the gymnosperms. It is surely not the only reason, though, because the phenomenally successful grasses are wind pollinated.

The flower parts that surround the seed (or seeds) often develop into an edible fruit or nut. Here again the feeding habits of animals are turned to the service of the plants. Squirrels carry off nuts to bury them for future use and often fail to dig them up again, thereby planting trees. Other mammals and birds eat fruits, the seeds pass unscathed through their digestive tracts and are deposited some distance away from the parent plant, along with a supply of fertilizer provided by the animal's fecal material. Indeed, some plants (for example, the mistletoe) have become so dependent on animal dispersal that the seeds cannot sprout unless the seed coats have first been softened by exposure to an animal's digestive juices. Other fruits, like burrs and beggar ticks, develop hooks and spines by which they attach themselves to fur, feathers, or human clothing and so are carried to new areas. Some flowering plants depend on wind dispersion and often have the fruit modified for this—think of the parachutes of the dandelion or the winged seeds of the maple. With such a variety of means of dispersal, the flowering plants, the latest to evolve of the land plants, rapidly spread throughout the world.

It is a far cry from a filamentous green alga waving in a current of water to a sturdy pine high on a mountain slope. Obviously the change was not accomplished in a single step. The surest way of tracing an evolutionary progression is through the fossil record, but unfortunately the record is usually far from complete and is sometimes ambiguous. It can be pieced out by a study of primitive forms that still survive, but every plant and animal living today has an equally long evolutionary history, and it is not always easy to decide which traits have been inherited virtually unchanged from ancestral forms and which represent more recent modifications. With this caveat in mind, we can still trace the broad outlines of the evolution of the land plants.

Green algae were present in the Precambrian era, over 1600 million years ago, but the filamentous types do not appear in the fossil record until the Middle Devonian period, less than 400 million years ago, and the Chaetophoraceae, the ones that appear to be closest to the land plants, not until the Early Jurassic about 180 million years ago. But the fossil record shows that primitive land plants were present much earlier, in the Late Silurian. So here is a major gap in the history of the plants at a most crucial point, the beginning of terrestrial evolution. Moreover, the earliest fossils of land plants are mostly incomplete fragments—here a bit of stem, there a spore case—and it is not always easy to reconstruct from these an accurate picture of what the whole plant was like. The first terrestrial forms probably are not represented by fossils at all. Almost surely they were low growing and showed little tissue differentiation, branching, stemlike plants that differed from their aquatic algal ancestors chiefly in their ability to survive out of the water. It may be that the sporophyte and gameto- phyte stages were equally well developed and existed as sepa- rate plants, but there is as yet no good evidence of the presence of the gametophyte in the fossil record of the Silurian.

Simple, branching plants with rounded spore cases at the tips of the branches were present in the Late Silurian. They had neither roots nor leaves but did have a central core of elon- gated cells with thickened rings in their walls—the first indi- cation of a vascular system.

The present-day clubmosses (not to be confused with true mosses) and the horsetails seem to be relatively little changed descendants of some of the earliest land plants. Modern club- mosses are small, seldom more than a few inches high, but some Devonian members of the group (Lycopsida) grew to be quite large trees and formed extensive forests. Clubmosses have true roots and usually small leaves. They bear sporangia on the upper surfaces of special leaves usually arranged in a conelike structure. The horsetails have jointed, ringed, hollow stems bearing whorls of leaves. Modern forms are a foot or so tall, though again some of the fossil members of the group (Sphenopsida) were good-sized trees.

Also common in the Devonian forests were members of an- other group that is more closely allied to the seed plants—the **ferns** (Filicinae). They have well-developed roots, stems, and large leaves which are thought to have evolved by modifications of the tips of the branches. These leaves differ from the leaves of the clubmosses and horsetails, which develop as outpocketings of the outer tissues of the stem. Both kinds increase the photosynthetic surface available to the plant; these are simply two different ways of doing the same thing.

The conspicuous ferns so often found growing in moist and shady places are sporophytes. But the ferns differ from the more advanced seed plants in having independent, free-living gametophytes. The little, hard, brownish dots you can see on the undersurface of a fern frond are clusters of sporangia. Usually ferns are **homosporous;** that is, they produce only a single kind of spore rather than having megaspores and microspores as the pine tree does. In dry weather the sporangia break open and the haploid spores are scattered by the wind. One that lands on moist soil in a cool, shady spot soon grows into a flat, heart-shaped little plant, a **prothallus.** It has chlorophyll and so is able to carry on photosynthesis, and from its undersurface fine, rootlike processes extend down into the soil to absorb water and minerals. This is the gametophyte of the fern. **Antheridia** (male sex organs) and **archegonia** (female sex organs) develop on its undersurface. When a mature antheridium is moistened, it releases sperm which swim through a film of surface water to the archegonia and there fertilize the eggs. The diploid zygote divides to give rise to the young fern sporophyte. At first it grows as an embryo, getting food and water from the gametophyte, but soon it develops roots and leaves of its own and takes up an independent existence. Then the gametophyte dies. Small and inconspicuous as the fern prothallus is, it is clearly a separate plant. Alternation of generations is thus much more evident in the ferns than it is in the seed plants.

Reproduction in the clubmosses and horsetails is very similar to that in the ferns. Sometimes the plant is heterosporous: it produces two different kinds of spores, two different gametophytes. Sometimes the gametophyte generation is more reduced. But in all of the more primitive land plants the sperm must swim through water to reach the egg. Nor do these plants produce seeds that can lie dormant in the soil during long periods of drought. They proved less successful as inhabitants of dry land than the seed plants which evolved later from the ferns and largely replaced the earlier groups. Yet it is only by comparison that the clubmosses, horsetails, and ferns seem not well adapted to terrestrial life. As first they were remarkably successful. Once established on land, they spread rapidly and in the Devonian period formed extensive forests. The land was prepared for the invasion of the animals.

SELECTED REFERENCES

ALEXOPOULOS, C. J., and H. C. BOLD. *Algae and Fungae.* New York, Macmillan Publishing Co., Inc., 1967.

DELAVORYAS, THEODORE. *Plant Diversification.* New York, Holt, Rinehart and Winston, Inc., 1966.

MULLER, W. H. *Botany: A Functional Approach,* 2nd ed. New York, Macmillan Publishing Co., Inc., 1969.

TORREY, J. G. *Development in Flowering Plants.* New York, Macmillan Publishing Co., Inc., 1967.

CHAPTER 3 The Arthropods

THE NEWLY GREENED earth was ready for the animals. And the animals were quick to follow the plants ashore. Indeed, the oldest fossils known that might have been terrestrial animals are from the Late Silurian and are therefore the same age as the earliest known fossil land plants. One of the animals that left these remains was a primitive scorpion, a representative of the arthropods. Almost certainly it was not the first animal to invade the land. It was a carnivore like all its close relatives, both its ancestors and descendants. It could not have fed directly on the plants. Another arthropod, a millipede or thousand-legger, has also been reported from Late Silurian time. Modern millipedes feed largely on decaying plant material and probably their ancestors did, too. It may be, then, that the first true terrestrial animals were precursors of the millipedes. Many people believe that both of these Silurian animals were still aquatic, but in the Devonian the arthropods were surely established on land.

If numerical superiority is a criterion of success, the arthropods are far and away the most successful of all the great divisions (called **phyla**) of the **animal kingdom.** There are more different kinds of arthropods (including crayfish, crabs, spiders, mites, ticks, insects, and so on) than of all other animals combined. Of these, the great majority are insects. Indeed, about three-fourths of the animal species living today are insects. They are all terrestrial or secondarily aquatic. They are found from above the Arctic Circle to the Antarctic, from underground caves to 20,000 feet up in the Himalaya Mountains, from ponds and streams to the driest desert. Some live part or all of their lives as parasites of other animals. No other class of animals has succeeded in occupying such a wide variety of habitats. Obviously the insects are superbly adapted to life on land.

Among the commonest and best known insects are the grasshoppers (Fig. 3-1). The grasshopper's body is divided into

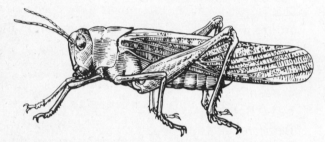

Fig. 3-1. A grasshopper, *Melanoplus spretus.*

three parts. There is a **head,** which bears the eyes; the powerful chewing jaws; and a pair of slender, **jointed appendages,** the "feelers" or **antennae.** The head is broadly joined to the **thorax,** the locomotor region of the body. It has three pairs of legs, the hind pair much longer and more muscular than the others, and two pairs of wings. The leathery forewings protect the more delicate, membranous hindwings. Both pairs are provided with a strengthening framework of thickened, hollow tubes called **veins.** When the animal is not in flight, the wings are folded back over the third division of the body, the **abdomen.** This part of the body is obviously segmented and lacks appendages except at the tip, where the male has genital claspers and the female an ovipositor.

The entire body of the grasshopper is covered with a noncellular, outer covering, the **cuticle,** which is secreted by an underlying layer of cells. The cuticle is composed largely of a tough but flexible substance called **chitin.** Over most of the body it is strengthened by the deposition of other complex chemicals, which make it rigid. An animal completely enclosed in a rigid armor would be quite unable to move around, something that is essential for all terrestrial animals. The hardened areas of the grasshopper's cuticle are divided into plates, separated by flexible areas, the joints. You can see the joints most clearly on the legs and antennae, but they are also present on the body proper. (The name "Arthropoda" means joint-footed; all members of the phylum have similar jointed appendages.) The cuticle is referred to as an **external skeleton.** It is quite different structurally from the skeleton of a vertebrate, which contains living cells and is enclosed within other tissues of the body. But the two kinds of skeletons have the same functions. Both provide support and protection for the internal organs of the body, and both serve as places of muscle attachment. Contraction of the muscles exerts pull on the various parts of the skeleton so that the combined skeletomuscular system acts as a set of levers to make bodily movements possible. The external skeleton of an insect is covered by a waxy waterproof coating with peculiar properties. It prevents loss of water from the body to the air, but at the same time at least some insects are apparently able to absorb water through it from a humid atmosphere or from damp soil.

A row of small, round openings can be seen along each side of the body of the grasshopper. These are the **spiracles,** the openings into the air tubes (**tracheae**). These air tubes are formed by infoldings of the body wall and are lined by continuations of the chitinous cuticle. Their walls are strengthened by spiral thickenings that serve the same function as the cartilaginous rings in the human trachea; they guard the essential air passages from collapse. A short way back from each spiracle lies a valve by which the trachea can be opened or closed. The valves help direct air flow through the tracheal system and, like the closable stomata of leaves, reduce the danger of water loss. The air tubes branch and branch again. They also connect with longitudinal tracheae that run the length of the body and with transverse ones that join the tubes on the opposite side. The result is an interconnected system of air passageways that reach to all parts of the body. Connected with the tracheae are smaller tubes (**tracheoles**) that penetrate the tissues and so carry oxygen directly to the cells. Air is pumped in and out of the system by the contraction of muscles of the body wall.

Like plants, insects have two transport systems, the one for respiratory gases just described and another for carrying food to the body cells and nitrogenous wastes away from the cells. This is the **circulatory system.** Since the blood is not the major vehicle for the transport of oxygen, it lacks the oxygen-carrying molecule, hemoglobin, and is almost colorless. Nutrients and waste products are carried in solution. For most of its course, the blood is not enclosed in blood vessels but fills cavities surrounding the internal organs and spaces in the appendages. There is a single long blood vessel running between the digestive tract and the body wall of the back from near the tip of the abdomen up into the head. The part that lies in the abdomen is the heart, a muscular, pumping tube with openings (ostia) in its walls through which blood from the abdominal cavity enters. Valves in the ostia and in the heart keep the blood flowing in one direction—forward through the aorta, the narrow forward extension of the tube that passes through the thorax into the head. The aorta is open-ended; the blood flows out and is carried by currents back through the body cavities into the vicinity of the heart. An **open circulatory system** like this is slower, less efficient than one in which the blood remains enclosed in vessels, as in the vertebrates. The cells of an active animal need a constant supply of oxygen; they can tolerate more easily fluctuations in the supply of nutrients. Suffocation is a much quicker death than starvation. Since the grasshopper's oxygen needs are taken care of by its tracheal system, its rather sluggish circulatory system is still quite sufficient for other transport duties.

The **alimentary canal** of the grasshopper is a long tube running from the mouth to the anus at the tip of the abdomen. It has three parts: foregut, midgut, and hindgut. Like the linings of the tracheae, the linings of the foregut and hindgut are continuous with the chitinous outer cuticle. At the point where the midgut joins the hindgut, several long tubules open into the intestine. These are the Malpighian tubules, the kidneys of the grasshopper. The nitrogenous wastes of insects are excreted into the intestine mainly as **uric acid,** which is not toxic and nearly insoluble. Very little water is lost in the process—an important adaptation for animals that live in a dry environment.

The brain of the grasshopper is a mass of nerve cell bodies (a **ganglion**) lying in the head above the foregut. From it two nerve trunks pass, one on either side, down around the foregut to connect with another ganglion lying beneath the alimentary canal. Paired nerve trunks running backward from this along the floor of the body cavity connect a series of ganglia from which nerves radiate out to all parts of the body.

The sense organs of the grasshopper are varied and complex, and although they differ structurally from those of man, they convey similar kinds of information about the external environment. Most obvious are the two large **compound eyes,** one on either side of the head. Each compound eye is a package of very many individual, six-sided visual units. Each unit faces the outside in a slightly different direction from every other unit so that each is stimulated by a different part of the total visual scene. The picture formed by an eye constructed in this way is like a mosaic built of many small stones. It is fuzzier in outline than the image formed by the human eye. The grasshopper has no way of changing the focus of its eyes and probably cannot see clearly anything more than a couple of feet away. But compound eyes are superbly adapted for detecting movement across the field of vision by the successive stimulation of the various facets. They can warn the grasshopper of the approach of an enemy and may also help a flying insect avoid obstacles and land accurately. In addition to its compound eyes, the grasshopper has three very small, simple eyes on the front of its face. Each is like a single unit of a compound eye. They do not form images, and it is not certain just what their function is.

The grasshopper can hear as well as see, but its ears are not located on its head. On either side of the first abdominal segment is a thinned place in the cuticle, the tympanum, by which sound vibrations are transmitted to sensory receptors in an underlying air space which is an expanded part of a trachea. As is so often true of animals that can hear, grasshoppers can also produce sounds. They do this by stridulating, that is, by rubbing one part of the body against another. Males of many kinds of grasshoppers have a row of tiny knobs on the inner side of the hind leg which they draw across a thickened vein on the forewing much as a fiddler draws a bow across the string of a violin. As with birds, each different kind of grasshopper has a distinctive voice. It apparently attracts and stimulates the female during courtship, and since a female only responds to the call of a male of the same species, the voice of the grasshopper also prevents cross breeding between different species.

The grasshopper has many other sense organs besides those of sight and hearing. Usually these organs are scattered over the body and the appendages, especially the antennae, and appear as minute hairs or spines or thin-walled cones or pegs. With them, the grasshopper smells and tastes and feels. It can find its food and sense vibrations on the surface on which it is resting, air currents on its face, changes in humidity.

Grasshoppers mate in the late summer; then the fields may resound to the shrilling of the males. When a receptive female responds to a male, he grasps her and injects a drop of seminal fluid containing the sperm into a special receptable near the tip of her abdomen, the copulatory pouch. The eggs are formed in her ovaries and each is provided with a protective shell. As they are being laid, they pass the opening of the copulatory pouch and the seminal fluid is shed on them. Since the shell forms around the egg before fertilization takes place, it is provided with a tiny opening through which the sperm can enter—reminiscent of the micropyle in the ovule of the pine, having the same name, and serving the same function.

Grasshopper eggs are laid in oval masses provided with a tough protective covering. Females of some species dig holes in the ground with their ovipositors and bury the eggs; others may excavate burrows in fence posts or fallen logs. The adults die when cold weather comes and only the eggs survive through the winter. The young that hatch the following spring resemble the adults in body form, though they are much smaller and do not have wings.

The hard, nonliving, external skeleton of the grasshopper provides protection against desiccation to the living organism within, much as the bark of the pine does, and like the bark it cannot grow or stretch to allow for increase in size. The bark of the pine flakes off; the skeleton of the grasshopper is shed all in a piece, including the linings of the foregut, hindgut, and tracheae. From time to time a new, soft cuticle forms under the old one, which then splits down the back, allowing the young grasshopper to work its way out. The new cuticle is elastic and stretches and the body increases in size. But the cuticle soon hardens and growth is again checked. With each successive shedding of the skeleton (**molting**) rudiments of the wings grow larger, but only after the last molt, when the grasshopper becomes adult, does it have fully formed wings that are capable of flight.

Nearly 1 million different kinds of insects are known, and they show a wonderful diversity of form; yet they all share many of the characteristics of the grasshopper. They all have the body divided into head, thorax, and abdomen; they all have external, nonliving skeletons; they all circulate air through a system of tracheal tubes. The adults have three pairs of legs on the thorax and usually two pairs of wings.

It may be that the wings of insects are their single most important feature, the characteristic that has allowed them to become the dominant group of animals on the earth today. With them, insects are able to escape from many of their predators; they can wander widely in search of food and mates; and they have been able to colonize the remotest islands. The most primitive living insects do not have wings; apparently they are the little modified descendants of the ancestral insect stock in which wings had not yet evolved. Among them are the silverfish (members of the order Thysanura), those notorious library pests that are able to feast on paste and the glue used in the binding of books.

The wings of the flying vertebrates, the birds, bats, and those extinct flying reptiles, the pterosaurs, evolved through modifications of appendages already present in their earthbound ancestors, but the wings of insects were entirely new structures. They developed from flaplike outgrowths of the body wall of the thorax; tracheae extending into these flaps were modified to form the veins of the wings. It is highly unlikely that when these flaps first evolved they could be used as wings, but what their original function was is still unknown.

The earliest known winged insects appeared in the Carboniferous period, about 300 million years ago. At first the wings could only be moved up and down and were held outspread even when the insect was at rest. The modern dragonflies (order Odonata) and mayflies (order Ephemeroptera) still have the primitive type of wings. Later the joint at the base of the wing evolved so that the wings could be folded back over the abdomen. This allowed the insects to invade even more environments, to creep under rocks and flakes of bark, to burrow in the trunks of trees or underground. In beetles (order Coleoptera) the front wings form a horny covering for the delicate, membranous hindwings. The beautiful colors and patterns of the wings of butterflies and moths (order Lepidoptera) are produced by a coating of tiny scales. Some insects, for example the lice (order Anoplura) and fleas (order Siphonoptera), have secondarily lost the wings, but they descended from winged ancestors.

The three pairs of appendages that form the mouthparts of insects show much variation (see Fig. 3-2). The primitive

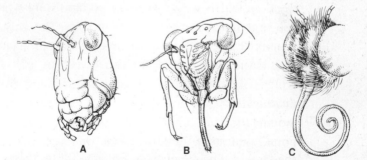

Fig. 3-2. Modifications of appendages for different feeding habits in insects. (A) Chewing mouthparts of a grasshopper; (B) piercing mouthparts of a bug (hemipteran); and (C) sucking mouthparts of a butterfly.

insect jaws were adapted for chewing, and many modern insects still chew their food. They may feed on plants, as the grasshoppers do, or on other animals, as the insect-eating dragonflies do. Butterflies have the mouthparts modified to form a long, slender tube for sucking nectar from flowers. When not in use the tube is coiled out of the way. Still other insects are able to both pierce and suck. From the human point of view they constitute some of the most destructive and dangerous of pests. Some, including the aphids and whiteflies (order Homoptera), suck the juices of plants. Others, among them mosquitoes, black flies, and midges (order Diptera) and lice and fleas, feed on the blood of animals, including man. They are more than nuisances; they transmit deadly diseases. Mosquitoes carry malaria and yellow fever; tsetse flies, African sleeping sickness; fleas, bubonic plague; lice, typhus. They have been responsible for more human death and misery than all the wars of history.

The newly hatched young of grasshoppers are like the adults not only in appearance, but also in habits. They live in the same places, eat the same food. And with each succeeding molt, they resemble the adults more closely. This pattern of development is called **gradual metamorphosis;** the young are called **nymphs.** In most other insects, the way of life of the immature differs from that of the adult. The adult dragonfly is a strong-flying predator, capturing its insect food on the wing. The immatures are also predatory but are secondary aquatics, lurking in hiding on the bottom of lakes, ponds, or streams.

Adaptation to life in the water has entailed modifications in the structure of the nymph. For one thing, there is the problem of obtaining oxygen. The tracheal system still transports air, but the spiracles are closed and the hind end of the intestine is somewhat enlarged and furnished with an abundant supply of tracheae. It functions as a kind of gill. Water is alternately drawn in and forced out through the anal opening, and there is exchange by diffusion of gases between the air dissolved in the water and the air in the tracheae. In spite of the differences imposed by a different mode of life, dragonfly nymphs still are not completely unlike the adults. The wing rudiments develop externally. Before the last molt, the nymph climbs out of the water, the skin splits down the back, and the fully winged adult emerges.

The immatures of many other insects are unlike the adults, not just in their way of life but in body shape and structure. Butterfly eggs hatch into caterpillars, fly eggs into maggots, beetle eggs into grubs. Young insects such as these are called **larvae.** The larva feeds and grows larger with each succeeding molt, but still does not look like the adult. Within its body, though, are clusters of cells, called **imaginal buds,** that play no part in the life of the larva. At last it **pupates;** that is, it goes into an inactive state. The larva of a moth spins a protective cocoon around itself; that of a butterfly pupates in its last larval skin. The pupal case of a butterfly hanging from a twig appears completely inert, but within the case most remarkable changes are taking place. The cells of many of the larval organs die and disintegrate and the materials of which they were formed provide nourishment for the cells of the imaginal buds. The latter begin to divide and develop into the organs of the adult. Then at last the insect undergoes its final molt and emerges from the pupal case, an entirely different animal. This type of life history is called **complete metamorphosis.** It allows the immature and adult to exploit different sets of environmental factors, to live in different habitats, to feed on different things. "A butterfly is a caterpillar's way of producing another caterpillar." Sometimes, as in the mayflies, the adults do not eat at all and survive only long enough to produce the eggs from which the next generation of larvae hatch.

The insects are the most successful, the most widespread, the most spectacular, but they are not the only terrestrial arthropods. The aquatic ancestor of all the arthropods is quite unknown. It is thought to have been a soft-bodied, wormlike creature, composed of a series of essentially similar segments called **metameres.** Each metamere bore a pair of leglike appendages. Evolution of the various arthropod lines from this primitive stock involved largely the development of a hard, jointed, external skeleton, the fusion of various segments of the body to form distinct regions, and the modification of the appendages for different functions. The first few metameres joined to form the arthropod head; the pattern of fusion of the metameres behind the head region varies from one class of arthropods to another.

Four major lines evolved from the ancestral arthropods. One comprised the trilobites, a group of primitive, aquatic arthropods. Judging by the fossil record, they were the most abundant animals in the early seas, but they have long been extinct. In them the body was divided into a head and a trunk consisting of from two to over forty metameres. The first appendages of the head were modified to form antennae-like structures. The appendages of the other metameres of the head and of the trunk were all quite similar and leglike.

The second line of arthropods, the Crustacea, probably evolved from primitive trilobites. They were present in the Cambrian period and almost surely first appeared before that. The head and body are fused, the abdomen separate. The first two pairs of head appendages are antennae; the last three act as jaws. The crustaceans comprise the largest living group of primarily aquatic arthropods: the crabs, lobsters, barnacles, and so forth.

The insects and their allies, the millipedes and centipedes, are sometimes lumped with the crustaceans as Mandibulata because they also have three pairs of head appendages modified to form jaws. Probably, though, they evolved separately from trilobite or pre-trilobite ancestors.

The centipedes, or hundred-leggers, have the body divided into a head region and a trunk region that, in different species, may have from as few as 17 to more than 190 segments. Since most of the segments have pairs of legs, obviously some centipedes have fewer than a hundred legs and some have many more. The first pair of trunk appendages comprises large claws connected to poison glands. Centipedes use them to subdue their prey of insects, worms, and other small animals. A centipede can also inflict a painful bite on anyone who handles it incautiously. Most centipedes are too small to be dangerous, but some tropical forms may reach a foot in length; a bite from such an animal can be really serious. Centipedes are found on all continents and most islands of the world, from sea level to above timberline, from the Arctic to the tropics, from humid forests to deserts. They are never as numerous as insects, perhaps because they never developed wings, but they are obviously well adapted to life on land.

The millipedes, or thousand-leggers, look something like centipedes but the resemblances between them are superficial. Millipedes also have a head and a trunk formed of a number of segments. Except for the first four and last one or two, the trunk segments are fused in pairs so that each apparent segment represents two metameres and each has two pairs of legs. Thousand-leggers are inoffensive animals, feeding mostly on decaying wood. They lack the poison claws of the centipedes, though most have glands that emit a repellent gas. Other than this, their only method of defense is to curl up into a ball. Like the centipedes, they are found on all continents.

Centipedes and millipedes, along with a couple of smaller, little known groups, are sometimes lumped together in the class Myriapoda (meaning "many-footed") but this does not seem to be a natural grouping. The centipedes, in fact, are probably more closely related to the insects than they are to the millipedes. The whole question of the relationships of these groups is very much up in the air, and the fossil record offers little help. "A fossilized myriapod or insect is the result of an incredible series of improbable circumstances whereas the fossilization of aquatic forms is merely unlikely" (Fox and Fox, 1964). Do they represent a single invasion of the land, or two? The earliest insects appear in the fossil record in the Upper Devonian. If they evolved from a terrestrial ancestor that also gave rise to the millipedes, that ancestor must have been present a good bit earlier. It is highly unlikely that they evolved directly from the millipedes. A millipede-like animal has been reported from the Upper Silurian. If indeed the millipedes were present that early, then almost surely the divergence between the line leading to the insects and the line leading to the millipedes occurred before the arthropods came ashore.

Whether or not the insects and the millipedes evolved from separate aquatic ancestors, there undoubtedly was a successful invasion of the land by another arthropod stock. The spiders (Fig. 3-3), scorpions, daddy longlegs, ticks, and mites all belong to the class Arachnida, representing the fourth of the major arthropod lines, the Chelicerata. These animals have two body regions. The head and the locomotor region of the body, corresponding to the thorax of the insects, join to form a **cephalothorax.** There are never any antennae; the first pair of appendages, in front of the mouth, is modified to form grasping claws used in feeding. They are called **chelicerae,** hence the name of the group. There are five pairs of appendages on the cephalothorax behind the mouth. The first of these pairs, usually called pedipalps, may be used for grasping prey, for chewing, for feeling, or for walking. The other four

Fig. 3-3. A spider, *Lycosa helluo*.

pairs are walking legs, though they may have additional functions. The abdomen is sometimes long and obviously segmented, but usually the abdominal metameres are fused to
form a round or oval unit with no apparent segmentation.
Sometimes, indeed, the abdomen is so closely joined to the
cephalothorax that no separate body regions can be made out.
The abdomen lacks walking legs in all arachnids but may have
appendages modified for other functions.

Though far overshadowed by the insects, if number of
species is a criterion of success, then the arachnids must be
considered the second most successful of the animal groups
that invaded the land. More than 50,000 species have been
described so far, and since many of them are minute to microscopic in size, it is sometimes estimated that as many more
species remain to be discovered. In contrast, only about 20,000
kinds of land vertebrates are known.

Best known of the arachnids are the spiders. They are widespread on all continents except Antarctica; most are large
enough to be easily seen; and the webs many of them weave

are even more conspicuous. Spiders usually have a rounded or oval abdomen that lacks any sign of segmentation and is joined to the cephalothorax by a slender stalk. All spiders are predators and have chelicerae armed with poison fangs with which they kill their prey. Not all trap their victims in a web; some are active hunters, some lurk in ambush. They are indiscriminate feeders, eating anything that they are large enough to subdue, including small birds and mammals, lizards and snakes, as well as other invertebrates. The female spider usually eats the male after copulating with him. Though deadly as a mate, she is generally a good mother, wrapping her eggs in a silken cocoon, guarding and feeding her young, sometimes carrying them on her back.

Probably the best known character of spiders is their spinning ability. Spider silk is finer, yet stronger, than the silk of a silkworm. It is formed by abdominal glands and extruded through one to four pairs of abdominal appendages called spinnerets. Not all spiders spin webs, yet all let out a thread of silk as they run. It may have been used originally just for the protective cocoon that is spun around the eggs. But the spiders have evolved a multitude of other uses for this versatile substance. A spider can let out a silken thread to lower itself to the ground, or to haul heavy prey up into its net. A male spider can identify a dragline let out by a mature, unmated female and follow it to its source. Trapdoor spiders construct silk-lined burrows with hinged, silken lids plastered over with dirt. The spider lurks in the burrow, holding the lid half open to watch for the approach of food or foe. It can dart out to seize its prey or, at the approach of danger, drop shut the trapdoor, which effectively conceals the mouth of the burrow. Whether or not a spider spins a web, and the kind of web it makes, depends on the species. Some produce an irregular, three-dimensional structure, some a funnel, some a flat sheet. Most elaborate are the large, patterned webs of the orb-weavers.

We once had a fishing spider (**Dinopus**) living on one of our ceilings. It lurked in a corner during the day but at night came out to hunt. It wove a rectangular net which it held spread out with its front legs and threw over any insect that ventured within range. The bolas spider of South Africa spins out a line tipped with a sticky globule which it twirls in the air to entrap its prey. Silk also provides spiders with a unique method of dispersal. Young spiders climb up to exposed places and each emits a strand of silk which is caught by a breeze; the little spiders are lifted into the air and go ballooning off, sometimes for hundreds of miles. No wonder spiders are so widely distributed.

Spiders are universally, and for the most part undeservedly, feared. Few are large enough, or produce a potent enough venom, to be dangerous to man. They do no damage to crops or livestock or buildings; they do not transmit diseases; and while they are indiscriminate in their choice of food, they probably destroy more harmful than beneficial insects. Less feared, yet far more destructive, and potentially dangerous, are members of another group of arachnids, the mites and ticks. Mites are all tiny and some are small enough to be invisible to the naked eye. The bloodsucking ticks are somewhat larger and, especially when engorged, are easily seen. Ticks and mites have the abdomen broadly fused to the cephalothorax so that the division between the two is not evident; the mouth region with the chelicerae and pedipalps is set off from the rest of the body as a headlike structure called the capitulum. All ticks are external parasites of vertebrates, but the mites are more diversified in their way of life. Some are free-living predators or scavengers. Some are plant feeders, most unusual among the arachnids. Many are parasitic; most species of plants as well as animals play host to one form or another of mite. The direct damage they do is usually more annoying than dangerous, but both mites and ticks transmit a variety of serious diseases—typhus, tularemia, Q fever, to name a few.

The scorpions suffer from as bad a reputation as the spiders, though again the majority of species are not dangerous to man. They are neither so numerous nor so widely spread as the spiders and the mites and ticks. The abdomen is long, obviously segmented, broad in front, narrow and tail-like behind. The pedipalps are large grasping claws. The poison glands are not connected with the chelicerae but with a sharp pointed sting at the tip of the abdomen. Like the spiders, the scorpions are all predators, seizing their prey with their claws and swinging their abdomens forward over their heads to inject the poison.

There are several other kinds of arachnids, but only the daddy longlegs or harvestmen are at all familiar. Most of the others are tiny tropical forms.

One of the really striking things about the terrestrial arthropods is that they are all quite small. The largest was an ancient dragonfly that had a wing spread of three feet, but in body size no land arthropod has ever been any larger than a fairly small mammal. Aquatic arthropods may grow somewhat bigger. One fossil eurypterid (a member of the long extinct group from which the arachnids evolved) reached a length of nine feet. Probably small size is a limitation imposed by the possession of an external, nonliving skeleton that must be shed from time to time. Before each successive new skeleton hardens, it offers little support for the body. In the more buoyant water, this is not so much of a problem, but in the less dense air, support is a major function of the skeleton. Only the trees, with their cambium cells constantly adding new material to the supporting wood, and the vertebrates, with their internal, living, growing skeletons, have escaped the limitations on size in the terrestrial environment. Small size, in turn, places another restriction on the arthropods. There is no room in the head of an insect or a spider for a large and complex brain. The behavior of the arthropods is largely stereotyped, instinctive. They have little learning ability and they cannot solve problems by reasoning. Despite their numerical success, they are not the culmination of life on land. For that we must turn to the vertebrates.

SELECTED REFERENCES

Fox, R. M., and J. W. Fox. *Introduction to Comparative Entomology*. New York, Van Nostrand Reinhold Company, 1964.

Prosser, C. L., and F. A. Brown, Jr. *Comparative Animal Physiology,* 2nd ed. Philadelphia, W. B. Saunders Company, 1961.

Russell-Hunter, W. D. *A Biology of Higher Invertebrates*. New York, Macmillan Publishing Co., Inc., 1969.

Wilson, J. A. *Principles of Animal Physiology*. New York, Macmillan Publishing Co., Inc., 1972.

CHAPTER 4 Odds and Ends

BEFORE WE CONSIDER the vertebrates, though, we should look at a few other kinds of organisms that, while usually inconspicuous, are also part of the terrestrial community. Some are microscopic, others are limited in numbers and in distribution; most have not completely freed themselves from the aquatic environment.

The most important of these incompletely terrestrial organisms are the **bacteria.** They are microscopic, single-celled forms that are often classified with the plants but are really neither plants nor animals. They are found in the oceans, in fresh water, in soil, and in the bodies of plants and animals. The bacterial cell lacks many of the structures found in the cells of plants and animals. For example, the nuclear material is not set off from the rest of the cell by a nuclear membrane. Like plant cells, most bacteria have rigid cell walls that help them maintain osmotic equilibrium, but these walls differ chemically from those of plant cells. If a bacterial cell is round, it is called a coccus; if rod-shaped, a bacillus; if a curved rod, a vibrio; if spiral, a spirillum. Sometimes the cells do not separate after cell division but form clumps or chains. Bacteria are really aquatic organisms, but because they are so tiny they can live and multiply in the merest film of water. Some bacteria are able to adapt to unfavorable conditions by forming **endospores.** A tough, resistant coat develops around part of the cell content, including the nuclear material. The rest of the cell disintegrates, but the endospore can survive for years. It is resistant not only to desiccation but also to freezing, to boiling, and to disinfectants. It is so light that it can be blown about as dust and so be carried for many miles. It is not a reproductive structure, as the spores of plants are. For the endospore to become an active bacterium again, it must be restored to a favorable environment, which for the bacterium means, above all, the presence of water. Only active bacteria grow and divide to produce more bacteria.

Some bacteria are internal parasites. Perhaps the first terrestrial plants and animals brought their bacteria ashore with them, and these have evolved along with their hosts. As endoparasites they are still living in an aquatic environment provided by the tissue and cellular fluids of the host. Examples from man are the bacteria that cause plague, cholera, gonorrhea, and typhoid fever. Most parasitic bacteria do not form endospores. They must be transmitted from one host to another by some means that prevents desiccation en-route— direct contact, polluted food and water, the bites of sucking insects. An exception is the bacterium that causes tetanus (lockjaw). It forms endospores that can lie dormant in dry dust for many years, yet quicken to activity when introduced into the bloodstream through a cut in the skin.

We are most apt to be aware of the bacteria that cause diseases in us, our livestock, or our crops, yet they represent only a small fraction of the total number. Far more numerous and far more important to us are the bacteria that live in the soil. Many of them are **saprobes;** that is, they get their nourishment by breaking down the complex organic molecules that make up the bodies of dead plants and animals. In other words, they cause decay. All the large organic molecules that are characteristic of living things contain carbon. The decay organisms release this carbon; by their respiratory activities they return it to the air as carbon dioxide so that it is again available to the plants to build into organic compounds. If it were not for the saprobes, the carbon would remain locked in death, the available supply would long since have been depleted, the plants would have died of carbon dioxide starvation, and the animals with them.

Other elements essential for life are also freed for recycling by the soil bacteria. **Nitrogen** is a special case. It is a major constituent of all living things. And it is the most abundant element in the atmosphere. Indeed, about 80 per cent of the air is nitrogen (N_2). But neither plants nor animals are able to use nitrogen in this form. The saprobic bacteria that digest the organic remains of plants and animals excrete ammonia (NH_3) as a waste product. Other bacteria are able to oxidize ammonia to nitrite (NO_2) and still other convert nitrite to nitrate (NO_3). These reactions release the energy by which the bacteria are able to synthesize the organic compounds they need. But the by-product, nitrate, is readily taken up by plants and used by them to build the proteins and other organic molecules that then serve as the source of nitrogen for animals. Some soil bacteria use nitrate as their oxygen source in carrying on cellular respiration and release the nitrogen, which is then unavailable to the plants, but still others can convert free nitrogen to ammonia and thus add it to the system again. This is called **nitrogen fixation.** One very important group of bacteria do not fix nitrogen when living free in the soil. They are capable of invading the roots of members of the bean family (including peas, clover, and alfalfa) and stimulating the growth of swellings called root nodules. The plant cells in the nodule are filled with masses of bacteria, and somehow the two in combination are able to fix large amounts of nitrogen. This is why growing a crop of clover or alfalfa in a field enriches the soil by adding to the available nitrogen.

Obviously, neither plants nor animals could survive long on land without the activity of the soil bacteria. Similar bacteria are found in surface waters and bottom muds, and they play similar roles in the aquatic environment. They must have accompanied the green plants in their invasion of the land.

The **fungi** are also important decomposers of organic remains. Mushrooms (Fig. 4-1), puffballs, and shelf fungi are

Fig. 4-1. A snail, *Helix aspera*, on a toadstool, *Boletus erythropus*.

familiar examples, but the molds, mildews, and yeasts also belong here. Fungi are usually classified as plants and for a long time it was believed that they evolved from the algae by the loss of chlorophyll and consequently of the ability to manufacture their own food. Now many mycologists (those who study fungi) think it is more likely that the organisms classes as fungi are not all closely related and that many of them evolved from single-celled animals rather than from algae. Be that as it may, fungi are either saprobic or parasitic. (A few are predaceous, able to trap minute worms and protozoans.) Some are aquatic, but many grow in moist places on land. Some, like the yeasts, are single-celled, but most are not. The parts of a fungus that you see, the mushrooms, the shelf fungi growing on the bark of a fallen tree, the colored spots on a slice of moldy bread, are **fruiting bodies,** the spore-bearing structures. The main body of the fungus is hidden out of sight. It is called a **mycelium** and is made up of a tangle of long, fine, branching tubes called **hyphae.** The walls of the tubes may contain cellulose, like the walls of plant cells; or chitin, which is otherwise found only in animals; or occasionally both. The mycelium secretes enzymes that dissolve and digest the organic material in which it is growing and then absorbs it. Parasitic fungi develop projections from the hyphae that push into the cells of the host to absorb nutrients. A mycelium may be roughly circular in shape and have the fruiting bodies develop around the periphery to give rise to a "fairy circle," a ring of mushrooms that spring up suddenly on a lawn. The grass within a fairy circle is usually lush because the old hyphae at the center of the mycelium die and release nitrogen to the soil. If the ground in which a fairy circle is growing is left undisturbed, the circle will reappear year after year and increase in diameter. Some are known that have been going for over 400 years.

Most fungi are haploid and many produce two kinds of spores. Asexual spores are formed directly by mitotic divisions of haploid nuclei and usually appear earlier in the life cycle than the sexual spores. The latter develop following zygote formation and then meiosis. Usually the zygote is the only diploid stage in the life history. The sexual spores of mushrooms are born on the gills on the underside of the cap; those of the puffballs develop inside the ball. Fungal spores are minute and tough, and are produced in astronomical numbers. Usually they are spread by the wind; indeed, the air in most parts of the globe is laden with them. This is one reason fungal diseases of plants spread so readily and are often so hard to eradicate.

The algae, by and large, are poorly adapted to life on land, though a few of them may grow in damp places. But a number of species of algae form symbiotic associations with certain fungi (symbiosis means "living together"). These algal–fungal combinations are known as **lichens.** The body of the lichen is a closely woven mat of fungal hyphae with the algal cells enclosed in the upper layer. About 23,000 different kinds of lichens have been described, each a combination of a specific fungus and a specific alga. Some form flat, encrusting masses, some are leaflike, some bushy. In a suitably moist habitat, the alga is able to live independently of the fungus, but the fungus cannot survive in nature by itself. The algal cells divide and multiply within the fungus. From time to time small clusters of fungal hyphae, each containing a few algal cells, form like dust on the surface of the lichen and are blown away. If they settle in a suitable spot, they give rise to new lichens. Sometimes the fungus develops a typical fruiting body. The spores are disseminated and germinate, and if the developing hyphae encounter the right sort of algae, they engulf them and go ahead to develop into lichens. If not, they die.

The photosynthetic activity of the alga provides food for both partners, but whether or how the fungus benefits the alga is still a matter of debate. It has been suggested that the former absorbs minerals and water for both and/or that it protects the latter from desiccation. But while the algae are able to exist without the fungi, they would be very much more limited in distribution than they are without their partners. Lichens abound in places that few other organisms find habitable—in the Arctic and Antarctic, in deserts, on the bare rocks of mountain heights. (They are also abundant in tropical jungles.) In the Arctic, reindeer moss (a lichen) is a major source of food for the caribou. Lichens on rock surfaces begin the processes of soil formation. Living, they help decompose the rock; dead, their bodies provide organic material.

Other associations between fungi and plants are called **mycorrhizae.** Here the fungal hyphae surround and invade the roots of one of the higher plants such as the conifers and orchids. The fungi derive nourishment from the roots, but the plants seem to benefit at the same time. Many plants do not grow well unless the specific fungi with which they normally form mycorrhizal associations are present. Apparently they absorb some of the nutrients they need from the fungi. Orchid seeds in the wild do not germinate unless they contain fungal filaments.

The fossil record of the fungi gives us little information about when they first appeared on land. Obviously the free-living forms could not have survived before there was an accumulation of organic remains which they could decompose for food. They must have followed the green plants ashore. Undoubted fungal fossils are known from the Devonian period and mycorrhizae from the Carboniferous period.

One other group of plants established a foothold on land, though they never succeeded in freeing themselves entirely from a dependence on surface water. This group comprises the bryophytes: the mosses and their allies.

A single moss plant is an inconspicuous little thing that you would not be apt to notice by itself, but since mosses usually grow in mats containing many thousands of individuals, they are often conspicuous parts of a woodland scene. A haploid moss spore that lands in a suitably moist spot develops into a long, threadlike, branching structure called a **protonema.** From it tiny, rootlike **rhizoids** grow down into the soil. They are not true roots because they lack the conductive tissues found in the root of the vascular plants. But they are able to absorb minerals and water. Small stemlike projections with leaflike blades develop from the protonema. These are not true leaves and stems. The central cells of the stem may lack chlorophyll and be somewhat elongated but they do not form xylem and phloem. A cluster of rhizoids develops at the base of each little plant and the protonema disintegrates. The closely growing but separate plants are the moss that one usually sees.

In contrast to the vascular plants, the conspicuous, independent part of the life cycle of the moss is the haploid gametophyte. The sex organs are the antheridia, in which sperm are formed, and the archegonia, each of which produces a single egg. Some species have both on a single plant; in others the individual plant is either male or female. Antheridia are usually round or oval in shape; archegonia are flask-shaped. In both, the developing sex cells are inclosed in a protective jacket of cells that do not take part in gamete formation. When the ovum is mature, cells in the neck of the archegonial flask disintegrate to form a narrow canal. During a time of heavy dew or light rain, the antheridia burst and the sperm swim through the film of moisture to the archegonia. They are attracted by chemicals released by the neck canal cells and swim down the canals to reach the eggs. The fertilized egg develops into an embryo within the protection of the archegonium and breaks out as the diploid sporophyte plant. Since it contains chlorophyll, it is able to manufacture its own food, but its foot remains buried in the tissues of the parent gametophyte, from which it absorbs water and minerals. It never becomes an independent plant. A capsule develops at its tip in which spores are formed by meiosis. They are disseminated by the wind to start the cycle again.

The mosses have a number of the adaptations needed for life on land. They are covered outside by a waxy, waterproof cuticle secreted by the epidermal cells. The gametes and the embryo are protected from desiccation during development. But mosses also lack some of the major requirements for success in the terrestrial environment. They do not have true vascular tissues; instead, nutrients and water pass from cell to cell by diffusion. The whole plant can absorb water, but there are no true roots to push down into the soil and absorb water from underground sources when surface moisture is lacking. Above all, the sperm must swim through a film of water to reach the egg. A very few mosses have been able to adapt to desert conditions. Usually they appear as blackened encrustations on bare rock faces but can spring to sudden brief activity after one of the rare desert rainfalls. Most mosses, though, are restricted to very humid environments.

The fossil record of the mosses is very poor, partly because they lack the hard parts found in seed plants and partly because they usually grow in places where bacterial decay takes place rapidly so that they are destroyed before they can be fossilized. Because in them the haploid gametophyte is the dominant stage of the life cycle and the diploid sporophyte never achieves an independent existence, botanists believe they represent a distinct evolutionary line from the one that led to the seed plants. They probably arose from a different stock of green algae, but what that ancestral stock was and when they first appeared on land are still open questions.

Turning again to the animals, one of the most successful of the major groups is the **phylum Nematoda.** The nematodes (roundworms) are probably surpassed in number of kinds only by the insects. Yet few people have ever heard of them, much less seen one. Many of them live in the bottom muds of oceans and fresh waters or as internal parasites of all kinds of plants and animals. But many others are found living free on land. They are basically aquatic; the terrestrial ones are most abundant in damp, humus-rich soil (a spadeful of rich garden dirt may contain a million of them). Roundworms are slender, elongate animals, round in cross section, usually tapering at each end. The free-living forms are minute, mostly less than 1 millimeter ($\frac{1}{25}$ inch) long. The body is not divided into segments. The digestive tract, a straight tube running from one end to the other, is surrounded by a fluid-filled body cavity.

Some roundworms are herbivorous, some carnivorous; some feed on decaying plants and animals. They seem to have few of the adaptations needed for life on land. They have no special respiratory organs. Apparently exchange of respiratory gases takes place simply by diffusion through the body wall. They have no special circulatory systems. Food is digested in the intestine and passes into the surrounding fluid-filled cavity. Thrashing movements of the whole body churn the fluid and apparently help in distribution, but this is a less efficient method than circulation by a blood–vascular system. Probably it only succeeds at all because the animals are so small. Roundworms have no special organs of locomotion. They swim or wriggle through the soil by a lashing movement in which the whole body is flexed first one way, then the other. They usually have no obvious sense organs except for some projections that are sensitive to touch. Apparently they also respond to chemical stimuli.

Yet the nematodes are adapted to life on land in two ways. In the first place, they are covered with a thick, nonliving cuticle secreted, as in the arthropods, by the underlying cells. This cuticle, like the arthropod skeleton, is shed from time to time during growth. The reproductive pattern of the nematodes fulfills all the requirements for terrestrial life. Fertilization is internal. The male copulates with the female and injects the sperm, which crawl up her reproductive tract. After fertilization a chitinous shell forms around the egg. Thus both the gametes and the embryo are protected from desiccation. The ovum contains enough yolk to nourish the developing embryo until it hatches as a larva. Unlike an insect larva, it closely resembles the adult, although it must undergo several molts before it reaches adult size. The eggs are produced in incredible numbers. They are quite resistant to dryness, and both larvae and adults may encyst so that they can survive long periods of drought.

The terrestrial nematodes did not evolve their cuticle, internal fertilization, and shelled eggs as adaptations to life on land. They share these features with their more primitive aquatic relatives and, almost certainly, with their aquatic ancestors. Perhaps the cuticle first evolved as an aid to locomotion. Unlike the arthropod skeleton, it is not divided into separate plates each with its own muscles so that the parts of the body can be moved relative to each other. But it does provide a mechanism by which the shape of the whole body can be deformed by the pull of the body-wall muscles, which all run longitudinally. Locomotion by whole-body flexion is not nearly so efficient as locomotion by flexing parts of the body, but it probably provided the nematodes with greater mobility than their (unknown) ancestors had. Protection of the gametes and of the developing embryo are not so essential in water as they are on land, but they are still advantageous. In these two important respects, then, the nematodes were preadapted to invade the land. Once there, their failure to develop respiratory and circulatory systems, better sense organs, more adequate means of locomotion, prevented them from really taking advantage of their new world. They are confined to existence as minute organisms dwelling in the interstices of the soil, usually in damp places. They cannot be said to have really conquered the land.

One other major invertebrate group includes more different

species than the Vertebrata—the **phylum Mollusca.** It comprises such varied animals as clams and oysters, squids and octopuses, snails and slugs. The basic molluscan body plan is quite different from that of the other invertebrates. There is a head bearing sense organs. Extending back from the head is a muscular, flat-surfaced foot on which the animal creeps along. Humped up above the head–foot is the **visceral mass,** which contains the internal organs. Covering over the visceral mass is the **mantle,** which usually secretes a calcareous protective shell. The hind end of the mantle is hollowed out to form a cavity into which the digestive tract and the excretory and genital ducts open and which typically bears gills.

Only one class of mollusks need concern us because only it has members that succeeded in becoming terrestrial. This is the **class Gastropoda,** including the snails (see Fig. 4-1) and slugs. The gastropod shell is all in one piece instead of having two halves hinged together like the shell of an oyster. Usually it is coiled in a spiral. The visceral mass extends up into the shell and the head and foot can be withdrawn into it. Sometimes a protective plate, called the operculum, develops in the foot. When the animal is withdrawn, this fits snugly into the opening of the shell. In the various kinds of slugs the shell is very much reduced or absent entirely.

A remarkable thing happens during the development of a gastropod. The visceral mass and mantle are twisted around so that the mantle cavity opens above the head.

Most snails and slugs are herbivores, but some are scavengers and a few are carnivores. The mouth has chitinous jaws and also a tonguelike structure, the radula, armed with rows of teeth. It is used to rasp off bits of food and also as a sort of conveyor belt to carry the food particles back to the gut.

The most primitive gastropods are marine, and the majority of them still dwell in the sea. Others live in fresh water or on land. The most obvious terrestrial adaptation of the land snails is that they have lost the gills, and the mantle cavity has been converted into a lung with a highly vascular lining and a narrow opening to the outside. This is why they are called pulmonate (meaning "having lungs") snails. Blood vessels from the lung, carrying oxygenated blood, empty into a two-chambered muscular heart from which blood is pumped out through a series of arteries into spaces surrounding the internal organs and muscles. The head has two pairs of retractile tentacles, the posterior pair bearing well-developed eyes. Marine snails excrete their nitrogenous wastes as ammonia, land snails as almost solid masses of uric acid crystals.

Land snails are hermaphrodites; that is, a single animal produces both eggs and sperm. But cross fertilization takes place; two animals mate and exchange sperm so that the eggs of one are fertilized by the sperm of the other. A snail has a copulatory organ, a penis, and fertilization is internal. The eggs are provided with hard shells. Marine snails hatch as larvae and undergo metamorphosis, but the young of terrestrial snails resemble the adults. They do not need to shed the protective shell at intervals as they grow because it too grows by the addition of new material around the edge.

Terrestrial snails are largely nocturnal. The foot secretes a slimy mucus to aid in crawling, and often the most obvious sign of their presence, aside from the damage they do in a garden, is the glistening trails they leave. Snails are active in humid weather, but they can survive in deserts, often in great numbers, by aestivating. The animal pulls into its shell and closes the opening, usually by secreting a plug of material which soon hardens. So sealed up, it can remain quiescent for many months. Slugs are essentially snails that, in the course of evolution, have lost their shells. Why they should have abandoned a structure that would seem to be an obvious advantage on land is a moot question. Perhaps they save enough energy by not having to lug a shell around to make up for the loss of protection. They are restricted far more than the snails to humid places in which there is little danger of desiccation.

The oldest known fossil land snail is from the Carboniferous period. Here again we have animals that were preadapted for life on land. The marine ancestors of the present-day terrestrial snails had protective shells and hard mouthparts, they were able to move about, albeit rather slowly, and presumably they had more or less adequate circulatory systems. They evolved a greater number of strictly terrestrial adaptations than the nematodes did. Even so, they are less successful on land than either the arthropods or the vertebrates. When a snail is active, the head–foot region is extended and is exposed to desiccation. And its crawling method of locomotion is painfully slow. There are modes of terrestrial existence that are forever closed to the snails and slugs. They cannot, for example, actively pursue active prey.

Earthworms are also animals that are incompletely adapted to life on land. They are members of the **phylum Annelida,** the segmented worms. We tend to think of worms as terrestrial, but the majority of them occur in the sea and others in fresh water. Still several thousand species of earthworms are found burrowing in humid soils or in leaf litter throughout the globe. The earthworm body consists of a series of segments which are not fused to form separate body regions as in the insects. Internally the segments are separated from each other by sheets of tissue that run between the body wall and the digestive tract. Externally the earthworm is covered by a thin, flexible, nonliving cuticle. Each segment bears four pairs of bristles. The body wall has two layers of muscles, one in which the cells run circularly around the body and another in which the cells run longitudinally. When the circular muscles of a segment contract, this puts pressure on the fluid which fills the cavity between the body wall and the gut. Being a fluid, it is not easily compressed, but it does flow; the segment changes shape, becoming longer and thinner. Conversely, when the longitudinal muscles contract, the segment grows shorter and thicker. The bristles are withdrawn when the segment elongates and extended when it shortens, so as to catch in irregularities in the soil and provide temporary points of attachment. If you watch a worm crawling, you can see the alternate waves of contraction that move along the body, seeming to push the animal forward. Because it has both longitudinal and circular muscles, an earthworm can move in a more coordinated fashion than can a roundworm.

Earthworms are vegetarians, feeding mainly on plant detritus in the soil. They have no jaws. The pharynx, the anterior part of the digestive tract, acts as a suction pump to draw food in through the mouth. Farther down the tract is a thick-walled, muscular gizzard with a hard lining in which the food is ground up. Marine annelids excrete ammonia. So do the earthworms in part, but they also convert some of their nitrogenous wastes into the less toxic urea. Both forms of excretion require large amounts of water. Earthworms do not have respiratory systems. The skin is rich in blood vessels and gas exchange takes place over the whole body surface, which must be kept moist. Oxygen is transported by hemoglobin dissolved in the blood. The circulatory system is closed; that is, the blood is always enclosed in vessels and does not lie free in the body cavity. There is no true heart, but some of the vessels have strong muscular walls which contract to pump the blood through the system. Earthworms are hermaphrodites, but like the snails they do not fertilize their own eggs. Pairs form and each worm exchanges sperm with its partner. Fertilization is internal. A number of segments toward the anterior end are expanded to form a girdle around the worm called the clitellum. It is rich in glands, and when the eggs are laid it secretes a protective cocoon around them. Marine worms hatch as larvae, earthworms in the adult body form.

The major problem of terrestrial life that the earthworms failed to solve is that of water conservation. They did not develop enclosed respiratory surfaces; they lose water readily through their skins and need much water for excretion. They compensate in part by being able to survive the loss of 50 to 80 per cent of their normal body water; even so, they desiccate so rapidly when exposed to dry air that almost everyone has seen the shriveled-up bodies of earthworms trapped on a sidewalk after a rain has stopped.

Finally, there are those intriguing little animals, the onychophorans, or walking worms (peripatus and its allies). About a hundred species are found in moist places throughout the warm regions of the earth. They hide by day in decaying logs and crannies in the earth and come out at night to walk slowly along searching for their small insect prey, mates, or new hiding places. Their proper classification is a matter of lively debate. Some consider them to be the most primitive of the arthropods; others place them in a separate phylum, descended perhaps from the ancestral stock that gave rise to both the annelids and the arthropods. And indeed they show a nice blending of wormlike and arthropodlike characters. The body is covered with a thin, wrinkled, flexible cuticle which is not divided into plates as the arthropod skeleton is. Externally the only sign of segmentation is the repeated pairs of clawed walking legs, from 14 to 44 depending on the species. But the arrangement of the internal organs shows signs of annelidlike segmentation. The head bears a pair of short, thick, antennae; eyes; jaws like short legs with enlarged claws; and a pair of leglike structures that can eject a milky fluid. This hardens at once into a tangle of sticky threads that can immobilize an attacking insect predator. Like the insects, walking worms have an open circulatory system and excrete uric acid crystals. They also have tracheae for gas exchange but differ from the insects in that the tracheae do not branch and join to form a connected system. Instead, each segment has a number of short, unbranched tubes opening separately to the outside. The tubes cannot be closed as the insect tracheae can, and as a result the walking worm cannot control water loss through them. This is probably the main reason onychophorans are restricted to moist habitats. The sexes are separate and fertilization is internal. Some species lay shelled eggs; others keep the eggs in the oviduct until the young hatch as miniature adults.

A fossil from a marine deposit of the Cambrian period appears to be that of an animal remarkably like a walking worm. All onychophorans today are terrestrial, but no evidence has as yet been found in the fossil record to tell us when they became so. In the absence of evidence, speculation can run free; it has been suggested that they were indeed the first land animals. Some entomologists believe that insects and their allies did not evolve from either crustacean or trilobite ancestors, but from the onychophorans. This may be one of the things we will never know.

A few other animals are found in moist places on land, such as the pill bugs and land crabs (crustaceans) and terrestrial leeches (annelids). But the story is always the same. Like the animals discussed in this chapter, they all failed in one way or another to complete the transition.

SELECTED REFERENCES

ALEXOPOULOS, C. J., and H. C. BOLD. *Algae and Fungae.* New York, Macmillan Publishing Co., Inc., 1967.

DELAVORYAS, THEODORE. *Plant Diversification.* New York, Holt, Rinehart and Winston, Inc., 1966.

MULLER, W. H. *Botany: A Functional Approach,* 2nd ed. New York, Macmillan Publishing Co., Inc., 1969.

RUSSELL-HUNTER, W. D. *A Biology of Lower Invertebrates.* New York, Macmillan Publishing Co., Inc., 1968.

———. *A Biology of Higher Invertebrates.* New York, Macmillan Publishing Co., Inc., 1969.

CHAPTER 5 # The Vertebrates

OF ALL THE INVADERS of the land, the ones most interesting to us are the vertebrates because we ourselves are vertebrates. Had it not been for the first tentative steps onto land of our early amphibian ancestors, we would not be here. Man could only have evolved from a terrestrial vertebrate ancestor.

The large **phylum Chordata** comprises three subphyla. Two contain a small number of completely marine animals, the sea squirts and amphioxus, beloved of embryologists. The third is the **subphylum Vertebrata:** the **fishes, amphibians, reptiles, birds,** and **mammals.** All chordates show at some stage in their life history three common characteristics. One is the presence of a series of **pharyngeal slits** opening to the outside in the wall of the pharynx, the anterior part of the digestive tract. The gills of the fishlike vertebrates are borne on the walls of the slits, and it is here that gas exchange takes place. Only traces of the slits appear in the early embryos of the higher vertebrates, which breathe by means of lungs. A second chordate characteristic is a **hollow dorsal nerve cord.** The nerve cord of the annelids and arthropods is a solid bundle of nerve fibers running through the body cavity below the digestive tract. From it nerves pass out to the organs and muscles. In the chordates the fibers surround a central cavity and the cord lies above the digestive tract. The anterior end is expanded to form the brain. Just below the nerve cord is a stiffening rod, the **notochord,** the character from which the chordates derive their name. The notochord of the vertebrates is more or less replaced during embryonic or larval development by a series of cartilaginous or bony vertebrae. Each consists of one or more skeletal elements forming a basal block, the centrum, from which an arch extends up to surround and protect the nerve cord.

The notochord marks the first appearance of the **internal skeleton** that is such an important feature of the vertebrate body plan. It consists of a mass of large, closely packed, fluid-filled cells, the whole surrounded by a dense fibrous sheath. Cartilage and bone are the two tissues that make up the vertebrate skeleton. Each consists of living cells scattered through a nonliving matrix secreted by the cells. The matrix of cartilage is usually less rigid than that of bone. One most important difference between the internal, living skeleton of the vertebrates and the external, nonliving skeleton of the arthropods is that the former can grow. The animals are freed from the restrictions on size imposed on the arthropods, at least partly, by the periods during growth when they are deprived of skeletal support.

The earliest amphibians evolved from ancestral fishes in the Late Devonian. For a time, until nearly the end of the Paleozoic era, they ruled the lands as the first terrestrial vertebrates. Today they are a waning lot, largely replaced by their better adapted descendants, the reptiles, birds, and mammals. But enough of them survive to show us how amphibians are built and how they live.

There are approximately 2400 species of living amphibians, arranged in four orders: **Anura, Caudata, Trachystomata,** and **Apoda.**

ANURA

The order Anura comprises the familiar frogs and toads, the dominant group of living amphibians, with about 2000 living species (see Fig. 5-1). As the name implies (*a* means "with-

Fig. 5-1. A tree frog, *Hyla squirella*.

out"; *ura,* "tail"), they have no tail. The hind limbs are modified for hopping. Often there is a free-living aquatic larval stage, the tadpole. It differs markedly from the adult—a tadpole has a tail, lacks legs, and breathes by means of gills. Many people think that frogs and toads are two different kinds of animals. Actually all members of the order can properly be called frogs. "Toad" is a common name. It is usually applied to more terrestrial members of the order, but not always. The Surinam toad is fully aquatic, even as an adult.

CAUDATA

The order Caudata comprises those amphibians familiarly known as salamanders or newts (see Fig. 5-2). They have tails and two pairs of legs, though the latter are sometimes very small. There are about 300 living species. Usually they have aquatic larvae with gills. These larvae resemble the adults in

Fig. 5-2. A salamander, *Plethodon glutinosus*.

body shape, and the major change at metamorphosis is the loss of the gills and a switchover to respiration through the lungs and/or skin. Sometimes there is no metamorphosis and the adults remain in the water as permanent larvae with gills.

TRACHYSTOMATA

The three living species remaining in the decadent group Trachystomata are restricted to the southeastern United States and northeastern Mexico. The sirens are aquatic, living their lives as permanent larvae that never metamorphose and come out on land. They have lungs, but they also have gill slits and gills, which are the primary respiratory organs. They have two tiny front legs but no hind legs. Many herpetologists consider them to belong to the same order as the salamanders, but they differ markedly from all known salamanders in a number of ways. For example, the sperm has a double tail instead of a single tail as in all other salamanders.

APODA

The order Apoda (meaning "without feet") is composed of the legless amphibians. Caecilians are slim, wormlike creatures 10 or 12 inches long. The adults are without gills or gill slits, but some caecilians have larvae that do have gills temporarily and that lead an aquatic existence before they metamorphose into terrestrial adults. Except for the gills, the larvae differ little from the adults. Caecilians are burrowing animals without any sign of limbs and with degenerate eyes. About 160 different kinds are found in forested areas throughout the tropics.

Structurally the amphibians show many modifications for a terrestrial existence that differentiate them from the fishes. Fishes are supported by the water in which they live, but when the amphibians moved out on land, the skeleton, while still serving for muscular attachment, took on the added function of providing support for the body. The most obvious skeletal difference between a fish and an amphibian is that the fish has fins, the amphibian has legs. Both the front and hind legs of all terrestrial vertebrates are built on a similar plan. You can identify the bones in your own arm or leg. Next to the body is a single long bone, the humerus in the arm, the femur in the leg. Joined to the lower end of the humerus are two more long bones, the radius and ulna, and to the lower end of the femur the tibia and fibula. Then come a number of small bones that form the wrist and ankle, the bones of the palm and foot, and finally the bones of the fingers and toes. The basic number of digits is five, and so this limb is known as a pentadactyl (meaning "five-fingered") limb. During the course of evolution, one or more of the digits may be lost, and various other bones in the limb may be fused or eliminated. Sometimes, indeed, one or both pairs of limbs disappear entirely, as in most snakes and the caecilians. All vertebrates that have pentadactyl limbs, or that are descended from ancestors that had such limbs, are classed together as **tetrapods** (meaning "four-footed"), in contrast to the vertebrates that have fins, the fishes. The tetrapods then comprise all terrestrial vertebrates and such secondary aquatics as whales and sea snakes.

The fins of fishes are used as steering devices, but the tetrapod limbs can support and move a body on land whereas the fins cannot. Both fins and limbs are attached to internal skeletal supports, the pectoral or shoulder girdle for the fore limbs, the pelvic or hip girdle for the hind limbs. The amphibian pelvic girdle is in turn attached to one of the vertebrae, which gives the girdle additional strength in its role of supporting the body. Animals living in the more complex environment of land need to be capable of more complex movements than do those living in water. The tetrapod limb is more maneuverable than the fishy fin. In addition, the first vertebra in the amphibian backbone is modified to form a movable joint with the skull. A frog or salamander can move its head up and down, a fish cannot.

Gas exchange in fishes takes place through the gills, in tetrapods usually through the lungs. These are essentially outpocketings of the pharynx to form enclosed respiratory surfaces through which gas exchange can take place with a minimum of evaporative water loss. Oddly enough, though, lungs evolved long before the first amphibian crawled out on land. They are found in some primitive fishes living today and were almost surely present in the aquatic ancestors of the tetrapods. Perhaps they developed first as accessory respiratory structures in fishes living in warm, muddy waters in which the oxygen content is low. If the fishes could not get enough oxygen from the water through their gills, they could come to the surface and gulp air.

In spite of their long evolutionary history, the lungs of present-day amphibians are not well enough developed to provide for all their respiratory needs. Some gas exchange takes place through the skin and through the lining of the mouth and pharynx. Frogs have better developed lungs than other amphibians, but some salamanders have lost the lungs entirely and rely wholly on the skin and lining of the mouth. The amphibian skin is thus an important respiratory organ. It is thin, well supplied with blood vessels, and provided with many glands, whose secretions keep it moist. This means that water is readily lost from its surface by evaporation—one major way in which the adaptation of the amphibians to life on land is far from complete. Most are restricted to places where water is readily available. A few frogs have invaded desert regions, but they spend most of their lives hidden away in protective crevices or burrows. Even so, the amphibian skin does show some adaptations to life on land. The outer layer consists of dead cells packed with a horny material called keratin. This layer, the **stratum corneum,** which is found in all tetrapods, helps protect the animal from abrasions. Since the cells of the stratum corneum are dead, the layer cannot grow and must be sloughed off from time to time. The underlying, living cells continue to divide. The older cells are pushed toward the surface, become keratinized, and die; so a new stratum corneum is being constantly formed. Some amphibians, like the leopard frog, shed the outer layer in flakes, in much the same way that the bark of a tree flakes off as the tree grows. Others, like the American toad, shed it in a single sheet as the insect sheds its cuticle. Remember, though, that the amphibian does not depend on its stratum corneum for support, nor does it have to wait for the underlying layer to harden. Shedding the skin is not nearly so traumatic as shedding the skeleton.

Many adult frogs, especially those that spend most of their

time away from water, have a special, noncellular layer of mucus embedded in the skin. It has a marked ability to bind water molecules and is probably an important defense against desiccation.

As usual in animals, excretion of nitrogenous wastes by the amphibians is linked to the problem of maintaining osmotic balance. It is further complicated by the fact that many species spend part of their lives as aquatic larvae, part as terrestrial adults. Freshwater animals need to eliminate the excess water that enters their bodies from the environment by osmosis; terrestrial animals must conserve water. Both the larvae and those amphibians that are aquatic as adults excrete copious amounts of very dilute urine, with most of the nitrogenous wastes in the form of ammonia. For those that become terrestrial as adults, there is a switchover at the time of metamorphosis to excretion of urea rather than ammonia. This means that the urine can be much more concentrated. Amphibians are able to regulate the rate at which urine is produced by the kidneys, and they have bladders in which urine can be stored and from which water can be resorbed if necessary.

Fishes have little in the way of a tongue—just a ridge of tissue on the floor of the mouth that can be moved slightly up and down but cannot be extruded. A terrestrial amphibian usually has a well-developed tongue that not only manipulates food around in the mouth but can also be shot out to capture prey. The food of fishes is wet when caught, but amphibians feed very largely on dry insects. They have glands in the mouth whose secretions moisten the food before it is swallowed.

Unlike the blood of insects, that of the amphibians must transport oxygen as well as nutrients, waste products, and hormones. They have a **closed circulatory system** in which the blood is pumped through vessels instead of simply flowing through the general body cavity. This is not really an adaptation to life on land, since fishes also have closed circulatory systems. Still it is far more efficient than an open system, and it is probable that without it the amphibians could never have become active terrestrial creatures.

The cornea, the outer covering of the eye, becomes opaque when it dries out. Amphibians have developed special tear glands around the eyes and movable eyelids that not only protect the delicate eyeball but serve to flush the fluid tears across it at intervals. Both are lacking in fishes.

Fishes are able to perceive currents in water through sense organs lying in a series of surface canals, the lateral-line canals. The precise function of these canals is still a matter of debate; probably the best suggestion is that they enable a fish to locate nearby objects in the water. Larval and aquatic adult amphibians still have lateral-line canals, but terrestrial amphibians lose them at metamorphosis.

Most fishes lack mechanisms for amplifying sound waves and hear very poorly, if at all. Frogs have such a mechanism. The hyomandibular bone, which in most fishes connects the upper jaw to the skull, now runs between the capsule housing the inner ear and a large ear drum (tympanic membrane) set in the side of the head. The bone is surrounded by the middle-ear cavity. Sound waves set the tympanic membrane vibrating, and the vibrations are transmitted by the hyomandibular bone (now called the columella or stapes) to a membrane-covered window in the inner ear, where the sensory cells are located. Amphibians also include the first vertebrates to have well-developed voices; not only do the frogs have voice boxes and vocal cords, the males also have resonating vocal sacs that open into the mouth. They probably have louder voices in proportion to their size than any other vertebrates. The calls of frogs are quite as distinctive as the songs of birds, with each species having its own particular call. They may vary from the chirping of the tiny grass frog, so high pitched that many people cannot hear it, to the deep-throated roar of a bullfrog. Frogs respond to the voice of members of their own species. The calling of the male at a breeding pond attracts others, both males and females; in a mixed chorus of several different kinds of frogs, the female approaches and mates only with a male of her own species.

There are fewer species of amphibians than of any other class of tetrapods (reptiles, 6000; birds, 8600; mammals, 3500). Yet they show by far the greatest diversity in reproductive habits. It is as if they were experimenting, trying this way or that to solve the problem of reproduction on land. The sirenids are aquatic in all stages of their life history, but the other groups of amphibians—the frogs, the salamanders, and the caecilians—all have members that reproduce on land.

Fertilization is internal among the caecilians and usually takes place on land; the male has a special copulatory organ for inserting the semen in the female reproductive tract. This solves the problem of protecting the sperm from desiccation. Most caecilians lay eggs in moist places on land. The eggs lack a protective shell and the embryos are susceptible to desiccation; the female may coil around them to protect them. Sometimes the young hatch as aquatic gilled larvae that make their way to water and pass through a long period of development before metamorphosing into terrestrial adults. Sometimes they metamorphose before they hatch and are terrestrial throughout life. Sometimes the eggs are retained in the uterus during development and the young metamorphose before they are born. The embryos may be nourished by oil droplets secreted by the wall of the uterus. This seems close to an ideal solution to the problems of terrestrial reproduction and indeed is very like that adopted by the highly successful mammals. But the blind, legless, burrowing caecilians in which it evolved were ill adapted in other ways to complete the conquest of the land.

The most primitive salamanders have external fertilization, and the eggs, larvae, and sometimes also the adults are aquatic. Fertilization is internal in other salamanders, though not by copulation. The male deposits spermatophores, small stalks capped with gelatinous packets containing sperm. By a more or less elaborate courtship he induces the female to take the sperm packets into her reproductive tract. The fertilized eggs are deposited some time later. They may be laid in water and hatch into aquatic larvae; they may be laid on land but still hatch into larvae that must make their way to water; or the young may metamorphose before they hatch. Again, though, the eggs lack shells and must be laid in moist places. A few species retain the eggs in the oviduct while development takes place. The female may give birth to aquatic larvae or to fully transformed young.

Frogs show an even greater diversity. Fertilization is usually external, with the male clasping the female and shedding sperm on the eggs as they are laid. Occasionally, though, it is internal, though not by spermatophores. The eggs may be laid in water, in a burrow in the bank, on a leaf above a stream, in a foam nest constructed by the female, under a protective log or rock on land. The eggs may be carried about in a pouch on the back of the female or in gelatinous strings entwined about the legs of the male. In one species the eggs are laid on land, the male watches them until the developing embryos can be seen moving around in them, and then he swallows them down into his vocal pouch. Here the eggs hatch and the young develop through metamorphosis and are "born" as fully formed frogs from the father's mouth. In another group of species the eggs again are laid on land, the father watches them until they hatch, the tadpoles wriggle their way up onto his back, and he carries them to water. One African toad with internal fertilization retains the eggs in the oviduct and the young develop through metamorphosis before birth.

Many modern amphibians, then, have adaptations that allow them to reproduce on land. In Paleozoic times, one group added further improvements and became the reptiles.

Characteristics of reptile reproduction include the following:

1. Internal fertilization.
2. Copulation, an effective method of achieving internal fertilization.
3. Eggs laid on land in a protected spot. (Sometimes the young are "born alive," but this is surely secondary in the reptiles.)
4. Direct development, without an intermediate larval stage.
5. Egg with a yolk mass (stored food supply) large enough to carry the embryo through development until the adult body form is reached.
6. Three extraembryonic membranes develop; one is the **amnion,** which surrounds the embryo with a fluid-filled sac.
7. A protective shell deposited around the developing egg by glands in the oviduct. This is only possible when internal fertilization occurs.

The first four characteristics are found in one group or another of modern amphibians. Eggs that are laid on land and that develop directly into the adult form have more yolk than eggs that are laid in water with the young hatching as larvae, though amphibian eggs never have as much yolk as do those of reptiles and birds. It is chiefly the presence of the extraembryonic membranes and shell that distinguish the developing reptile egg (such an egg is called **cleidoic**). All vertebrate embryos have a cellular membrane surrounding the yolk mass, with the embryo lying on top. In the reptiles, folds of tissue grow up around the embryo to meet and fuse above it, forming an outer membrane, the **chorion,** and an inner membrane, the amnion, which encloses the embryo proper in a protective, fluid-filled sac. The third membrane, the **allantois,** grows out from the hind end of the embryonic gut and pushes between the chorion and amnion. It serves as a storage place for waste products. Birds and mammals share these membranes with the reptiles, so the three groups are often called **amniotes.** Figure 5-3 shows a developing reptile embryo with its membranes.

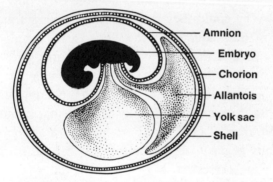

Fig. 5-3. The formation of the amniote egg.

It has been suggested that, as the amount of yolk increased in the terrestrial egg, the developing embryo tended to sink down in it so that the surrounding extraembryonic tissue was thrown up into folds that finally closed over the embryo, giving rise to the chorion and amnion. Another idea that has recently been proposed is that the first of the amniote extraembryonic membranes to evolve was the allantois. With eggs laid on land, water conservation becomes a problem. The allantois may have served as a storage sac for urine, from which water could be reabsorbed if necessary. As it developed, it pushed the folds of surrounding tissue around the embryo to form the chorion and amnion. At any rate, it is clear that the reptile egg, with the embryo protected both from desiccation and from shocks and jars by the amnion and shell, is far better adapted to life on land than is any amphibian egg. We call the first animal that evolved such an egg the first reptile.

It is an axiom of evolutionary thought that the structures and functions developed by an organism, while they may be pre-adaptive, must also be of immediate benefit in the situation in which the organism finds itself. What were the environmental pressures that led so many amphibians, including the ancestors of the reptiles, to take up the habit of laying their eggs on land? Why could they not just go on laying aquatic eggs that develop into aquatic larvae? For many amphibians, as for many insects, this is a successful pattern. One suggestion is that the ancestors of the amphibians with terrestrial repro-duction lived in regions where the ponds were subject to periodic drying. They developed eggs that could withstand a certain amount of desiccation, and more and more these am-phibians began to lay their eggs on the banks rather than in the water. Terrestrial eggs, then, evolved as a defense against desiccation. Another idea is based on the fact that aquatic eggs and larvae are subject to very heavy predation by other water-dwellers. By this hypothesis, the terrestrial egg was a defense against predation. A third idea is based on the distribution of the modern amphibians that reproduce on land. Almost in-variably they live in mountainous regions where quiet, open bodies of water are few and the usual aquatic habitat is a swift-flowing stream. Here sperm, eggs, and larvae are all in danger of being swept away by the current. Internal fertiliza-tion, terrestrial eggs, and direct development would all be ad-vantageous to an amphibian living in a region where the only available aquatic site was a raging mountain torrent. It will probably never be known which, if any, of these ideas is correct.

To ensure that at least some of the young will survive to adulthood, amphibians that lay eggs in water have to produce a great many more eggs than do those that lay their eggs in protected places on land. The female of a water-breeding toad may produce up to 25,000 eggs at a time, whereas the female of *Sminthillus limbatus* of Cuba, a small terrestrial breeder, lays but a single egg. This represents an enormous saving in the amount of energy that must be expended in reproduction.

Most animal tissues disintegrate rapidly after death and are almost never fossilized. Bone is an exception, and since vertebrates are, almost by definition, animals that have bone, we have a better fossil record of the evolution of the vertebrates than of almost any other group. There are tantalizing, frustrating gaps, but in general the story is clear.

Among the fishes that swarmed in the Devonian seas were members of a group called the **sarcopterygians,** or fleshy-finned fishes (*sarco* means "flesh"; *pterygium,* "fin"). In most fishes the paired fins are supported by parallel horny rays, but the sarcopterygian fin has a central bony axis fleshed out by muscles to form a lobe at the base. This type of fin was apparently the forerunner of the tetrapod limb. These fishes had lungs, as indeed most primitive fishes apparently did. In addition, they had passages running from the nasal cavities to openings in the roof of the mouth, the **internal nares** or **choanae.** The nasal cavities of other fishes are blind pockets with no connection to the mouth, but all tetrapods have internal nares. Few fleshy-finned fishes survive today—only the lung fishes and the ocean-dwelling coelacanth. But one line, the rhipidistians, before they died out at the end of the Paleozoic era, gave rise to the amphibians.

The ancestors of the amphibians apparently lived in shallow, warm, freshwater ponds. The oxygen content of such water is low; respiration through gills may have been insuffi-

cient; and the fishes may have come frequently to the surface
to take air into their lungs through their nasal passages. They
were carnivorous, feeding on other fishes. As they stalked their
prey through the thick growths of aquatic plants, they may
have used their well-developed muscular fins to propel themselves along the bottom. Why did they start to come out on
land? It may be that the ponds in which they lived dried up
from time to time and that some of them were able to heave
themselves out on land and cross to larger bodies of water.
They were the ones that survived and gave rise to the amphibians. Or it may be that the young were preyed on by the
adults and spent most of their time close to shore as the
young of many fish with similar habits do today. Perhaps they
first fled ashore in the rush to escape a pursuing adult. Gradu-
ally they spent more and more time on land and began feeding
on the terrestrial arthropods that had preceded them.

However it was, by the end of the Devonian period the first
tetrapods had appeared. Best known is *Ichthyostega,* an animal about 3 feet long with four short but sturdy legs. It had a
fishlike tail and lateral-line canals and probably still spent
much of its time in the water. Soon a number of amphibians
evolved that were well adapted to life on land. With no competitors, they flourished and spread widely.

But amphibian domination of the land was short-lived. By
Pennsylvanian times the reptiles had evolved from the amphibians and rapidly replaced them as the dominant class. We
have said that the first animal to lay an amniote egg was the
first reptile, but eggs are seldom preserved in fossil form (the
oldest known amniote egg is from the Permian). We base our
identification of fossil vertebrates on skeletal characteristics
and assume that by the time an animal had evolved a skeleton
more like that of a reptile than that of an amphibian, it had
also evolved the reptilian pattern of reproduction. There are

a number of transitional forms (we are, after all, dealing with a continuum), and there may always be room for debate as to just where the line between the fossil amphibians and reptiles should be drawn.

The amniote egg is not the only way in which reptiles surpass the amphibians in adaptation to life on land. The first and second vertebrae in the neck region are modified so that a reptile can turn its head from side to side as well as moving it up and down. The pelvic girdle is attached to two or more vertebrae instead of only one. In amphibians and primitive reptiles the limbs sprawl out to the sides, but in advanced reptiles the limbs are rotated and brought under the body. The knee points forward and the elbow backward, making for greater support of the body and more efficient locomotion.

The lungs of reptiles are better developed; the skin no longer is needed as a respiratory organ. Most fishes are covered with a protective layer of bony scales that develop in the lower layer of the skin (the dermis). Some ancient amphibians still had parts of the body covered with dermal scales, and traces of them can be found today in the most primitive caecilians. But the armor of dermal scales had to be sacrificed by the amphibians in order to allow the skin to be used in respiration. It is thin and easily torn, scant protection against the abrasive forces to which animals not cushioned by water are subject, and little help in the battle against desiccation. Reptiles evolved a new kind of scaly covering. The scales are horny rather than bony, and the outer layer of the skin is also involved in the formation of the scales. The scales of turtles and crocodilians are separate but those of lizards and snakes form as continuous sheets (see Fig. 5-4). Some reptiles have also redeveloped bony dermal plates. The shell of a turtle consists of dermal bones overlain by horny scales. Reptiles have few skin glands and lose little moisture through their horny outer

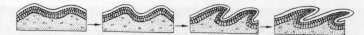

Fig. 5-4. The formation of snake scales.

covering. They are much more resistant to desiccation than the thin-skinned amphibians. Also they excrete much of their nitrogenous waste as uric acid. This is not only an important means of water conservation but is essential for animals that lay shelled eggs on land. The nitrogenous wastes produced by the developing embryo must be stored with it in the shell. Uric acid is nontoxic; ammonia and large quantities of urea are toxic. Not just the reptiles, but the other animals that lay shelled eggs on land—birds, snails, and insects—excrete uric acid. Mammals, though descended from reptiles, have given up the cleidoic egg in favor of retention of the developing embryo in the uterus of the mother and have returned to excreting urea.

The burgeoning of the reptiles was the most spectacular chapter in the history of terrestrial life. For 200 million years, from the Permian period to the end of the Cretaceous, reptiles ruled the land. They included the dinosaurs, some of them the largest animals that ever walked the earth. Some were slow-footed herbivores, some fast-moving bipedal carnivores that ran rapidly on strong, birdlike hind legs. Some reptiles, the pterosaurs, took to the air. This line culminated in the Cretaceous *Pteranodon,* which had a wing spread of 25 feet. Others returned to the seas to live like the modern whales and dolphins. But by the end of the Cretaceous period, the reign of the reptiles had come to a close. Only a few lines survived—the turtles, the crocodilians, the tuatara of New Zealand, and the one really successful group of modern reptiles, the lizards and snakes (see Fig. 5-5). Long before the great extinction, though, the birds and mammals, the dominant

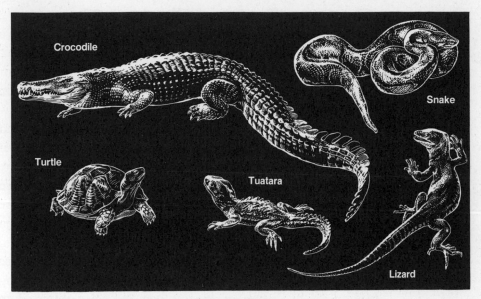

Fig. 5-5. Types of living reptiles.

tetrapods in the world today, had evolved from their reptilian ancestors.

There can be no doubt that the reptiles were eminently successful as terrestrial animals. Their long period of dominance, their great abundance and diversity, attest to this fact. Yet, judging by modern forms, there was one problem of life on land that they were not really successful in solving. That problem is posed by the fluctuating temperatures found in the terrestrial environment. Heat as a form of energy is a by-product of the chemical processes going on in an animal's body (its metabolism). When a reptile is at rest in cool surroundings, its metabolic rate is low and what heat it produces passes out of the body rapidly. Its internal temperature tends to be related to that of the surrounding air. To warm itself up enough so that it can become active, it must absorb heat either from the surface on which it is resting or from solar radiation.

Animals that are dependent on external sources of heat are said to be **ectothermic.** This does not mean that an active reptile has no control over its body temperature. It can maintain a relatively high and stable internal temperature for hours. By and large its control is behavioral. It basks in the sun to absorb heat, then moves into the shade or into a burrow if it gets too warm. Still the ectothermic amphibians and reptiles are limited in activity and in distribution by their dependence on external heat sources. When temperatures are low, they are sluggish or torpid and unable to escape from predators. In regions where the winters are severe, they are active only during the short summer months. Few have been able to invade the Arctic regions.

The birds and mammals are **endothermic;** they produce enough body heat so that they can maintain high and relatively constant body temperatures even in cold weather. Both groups have developed efficient insulating coverings: feathers in the birds, fur in the mammals. These can be fluffed out to trap a blanket of air next to the skin, which cuts down on the loss of body heat. They can regulate their metabolic rate to a considerable extent so that they produce more body heat in cold weather than in warm weather. Shivering is a series of muscular contractions that produce heat and serve to warm the body up. Birds and mammals are able to remain active over much longer periods of time, through much greater extremes of environmental temperatures, than can the reptiles. They are now the major groups of terrestrial vertebrates. The birds took to the air, modifying the basic tetrapod forelimbs into wings. The mammals, for the most part, remained on the ground and eventually gave rise to man, the single species that has claimed for itself dominion over the whole earth. Whether the earth will remain a suitable habitat for life may depend on how wisely man exercises that dominion.

SELECTED REFERENCES

COLBERT, E. H. *Evolution of the Vertebrates,* 2nd ed. New York, John Wiley & Sons, Inc., 1969.

GOIN, C. J., and O. B. GOIN. *Introduction to Herpetology,* 2nd ed. San Francisco, W. H. Freeman and Co., Publishers, 1971.

GORDON, M. S., et al. *Animal Physiology: Principles and Adaptations,* 2nd ed. New York, Macmillan Publishing Co., Inc., 1972.

OLSON, E. C. *Vertebrate Paleozoology.* New York, John Wiley & Sons, Inc., 1971.

ROMER, A. S *Vertebrate Paleontology,* 3rd ed. Chicago, University of Chicago Press, 1966.

CHAPTER 6 Muddy Waters

AS WE LOOK AT the evolutionary record, two things stand out rather clearly. One we have mentioned already: of the many major kinds of plants and animals that evolved in the water, only three were completely successful in conquering the land. The other is that, once the invasion of the terrestrial environment began, evolution took place very rapidly.

Each of the three groups underwent a major adaptive radiation, giving rise to many different kinds of organisms adapted to many different habitats and ecologic niches. The first land plants appeared at the end of the Silurian period, and by mid-Devonian times extensive stands of tree ferns and treelike club mosses were present. The arthropods were quick to follow the plants, and in the Devonian both spiderlike forms and winged insects had appeared. The earliest amphibians had evolved by Late Devonian time, a number of very distinct types were present in the Mississippian, and reptiles appeared in the Pennsylvanian. Almost all known fossiliferous deposits from the Carboniferous (Mississippian + Pennsylvanian) are coal measures formed in low-lying swamps. Dry-land deposits are usually formed in uplands and are more subject to subsequent erosion. As a consequence, they are less often preserved. The fossil record is thus biased in favor of the swamp dwellers, and indeed most of the known early tetrapods were aquatic or semiaquatic. Even so, some of the Pennsylvanian reptiles and amphibians seem to have been fully terrestrial animals, which suggests that dry-land plants and arthropods were also present.

When we turn from the question of what took place to the question of how it took place, the water becomes muddier. Even so, with recent advances in the study of **deoxyribonucleic acid (DNA)**, the picture of the mechanisms underlying these rapid evolutionary advances becomes a little less murky. What follows is highly speculative, and alternative hypotheses have been advanced to explain some of the facts. But it seems to us that the ideas expressed here best account for the similarity of evolutionary pattern found in three such disparate groups as the vascular plants, the arthropods, and the vertebrates.

When it was discovered that DNA is the substance of which **genes** are made, it became probable that there had been an overall increase in the amount of DNA per cell during the course of evolution. Genes determine what **proteins** the cell is able to form. Proteins make up a large part of the physical structure of any cell and they are also the **enzymes** that determine what chemical processes it is able to carry on. With increasing complexity of structure and function, more genes are needed to carry the increased information required to encode the organism. And indeed we do find that there has been an over-all increase in amount of DNA per cell with evolutionary advancement. Animals and plants have more DNA than bacteria, and complex multicellular organisms have more DNA than simple ones have. But this is not all there is to the story. There is frequently an enormous variation in the amount of DNA found in the members of any one group. The European fire-bellied toad has nearly ten times as much DNA as the American spadefoot toad, and the Congo eel, a salamander, has about sixty times as much. All salamanders and many frogs have more than man does. Similar variations are found among vascular plants and insects. Even among the mammals, which show much less variation than most classes, an aardvark has twice as much DNA as a free-tailed bat. It is simply not credible that it takes so much more DNA to encode a salamander than it does to encode a man, or that one toad is that much more complicated than another toad.

It has been calculated that in man perhaps 90 per cent or more of the DNA is not "genes" in the usual sense. It plays no role in the synthesis of proteins. It has been called "junk" DNA. Functionally, then, it is possible to divide DNA into two classes: operational and uncommitted. What we here term "operational" is simply the DNA that forms the genes that are used to structure the organism and keep it operating. **Uncommitted DNA** is not known to differ chemically from **operational DNA.** But it plays no part in determining the structure of the organism or in the day-by-day functional activities that keep it going. We believe it is this uncommitted DNA that accounts for much of the difference in amount of DNA in forms of a comparable degree of complexity. Some simply have more of it than others.

We also believe that it is the presence of large amounts of uncommitted DNA that permits the **major adaptive shifts** that have occurred from time to time in evolution. When a mutation occurs in a structural gene, one that encodes for a given protein, the chances are very great that the mutation will be either neutral or deleterious. Either it will not affect the functioning of the enzyme, or it will affect it for the worse. The enzyme may not work as efficiently, or it may not work at all. It is highly improbable that a single mutation would at one step produce a new enzyme capable of carrying on a different function that would be of immediate value to the organism. Even if it did, it would be at the expense of the previous function, which was also of value. Suppose, though, that the amount of nuclear DNA should be increased by introducing multiple copies of some or all of the genes. As long as one or more copies of the gene remained in good working order, those not needed for the operation of the organism would be free to mutate over and over. The chances would be very much improved that eventually some of them would be changed in such a way that they could encode for new enzymes without first sacrificing the old. Then the organism would be able to carry on new processes, to develop new structures. Perhaps a soft-bodied plant could now produce hard, supporting tissues, or an animal a shelled egg. It seems probable, then, that the the major adaptive shifts in evolution have been preceded by, and have been made possible by, major increases in the amount of DNA per nucleus.

The major differences in amount of DNA that we sometimes find between the species of a single modern genus indicate that changes can occur quite easily. They should warn us to be careful in extrapolating back to ancestral forms for which we cannot now measure the amount of DNA. It is also true that DNA values are known for only a small proportion of living species. Still, what evidence there is seems to indicate that in the vertebrates and plants, at least, evolution of terrestrialism did involve increases in the amount of DNA. The fish with the least amount of DNA yet recorded (a puffer) has about 0.8 picogram (10^{-12} gram), and the modal value for the teleosts, which make up the great bulk of the ray-finned fishes living today, is about 1.7 picograms. No rhipidistians have survived to the present, so we know nothing of the DNA content in the group that gave rise to the tetrapods. Of the other fleshy-finned fishes, the coelacanth has 6.5 picograms and the lungfishes parallel some of the salamanders in having fantastically elevated amounts (100 to 284 picograms). The modal value for the frogs, the best known group of amphibians, is about 8.4 picograms, and the least amount recorded for any terrestrial vertebrate is about 3 picograms. High values of DNA per cell thus seem to be characteristic both of the fleshy-finned fishes and of the tetrapods that evolved from them. The evidence is more scanty for the plants, but even so the primitive land plants seem to have more DNA than the green algae.

It is also obvious, though, that the amount of DNA is frequently reduced during the course of evolution. We usually find in any one lineage that the more highly specialized members have less DNA than the more primitive ones. In general, amphibians have more DNA than reptiles and reptiles more than birds. Gymnosperms have more than angiosperms. Among insects, the rather primitive grasshopper has almost a hundred times as much as the very specialized fruit fly.

On a short-term basis it is uneconomical for a plant or animal to carry a heavy load of excess DNA in its genome, to have to duplicate large amounts of genetic material each time cell division takes place. This is especially true if much of the genetic material is making no immediate contribution to the life of the organism. The first animals and plants that moved out onto the land were well endowed with uncommitted DNA. They were able not only to make this major adaptive shift but also to radiate rapidly into a variety of different habitats and to take up many different life styles. As they did so, they tended to lose excess DNA. At the same time they also lost much of their potential for radical modification. They became specialized for one or another way of life.

In more specialized groups, perhaps because their members lack the masking effect provided by the presence of multiple copies of many genes, minor genetic differences seem to be expressed more readily. A teleost fish can easily develop longer fin rays, brighter colors, spinier scales; it can become better adapted to this or that specific aquatic habitat. A basic teleost stock can give rise to a host of different species, but it cannot evolve into something other than a fish. The birds, which have less DNA than any other class of tetrapods, also include more species than the reptiles, mammals, or amphibians. A correlation between small amount of DNA and extensive speciation has been noticed in plants and insects as well as in the vertebrates. As a general rule, then, groups that are primitive or unspecialized, that is, have not departed greatly from the ancestral stock, have larger amounts of DNA and are represented by relatively fewer species. On the other hand, groups that are specialized tend to have small amounts of DNA and many species (see Fig. 6-1).

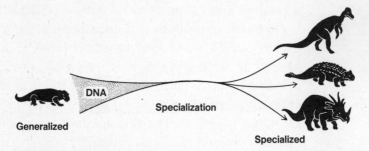

Fig. 6-1. Did the evolution of dinosaurs involve loss of DNA?

There are various ways in which changes in the amount of DNA can come about. One common form of increase is the establishment of **polyploidy.** Instead of the normal diploid condition, with two sets of chromosomes, a plant or animal can have three sets (**triploidy**) four sets (**tetraploidy**), or even higher numbers of sets. Polyploids can be formed in a number of ways. For example, sometimes normal meiosis fails to take place during the formation of the gametes. An egg may wind up as a diploid, with two sets of chromosomes instead of the usual one. If it is fertilized by a haploid sperm, the resulting offspring is triploid; if by a diploid sperm, the offspring is tetraploid.

To go into all the various mechanisms that result in polyploid formation would require a chapter in a genetics book and we will not attempt it here. Suffice it to say that polyploids are far more common in plants than in vertebrates. It has been estimated that more than a third of the modern species of flowering plants are of polyploid origin. Still, several polyploid species of amphibians have been found in recent years. Of two species of gray tree frogs in the United States, one is diploid and the other is tetraploid. No one has yet been able to figure out how to tell the two apart morphologically, but the breeding calls of the males differ and the females recognize and respond only to the calls of their own species. A South American frog is octoploid, with eight sets of chromosomes, and a number of lizard species are triploid. Probably as more species of animals are examined cytologically (most have not been as yet), more polyploids will be found. At any rate, it is clear that polyploidy can occur in vertebrates, and that it results in a marked change in the amount of DNA.

Not all changes in DNA amount can be attributed to polyploidy. Sometimes during meiosis the members of one or more of the chromosome pairs fail to separate and gametes are formed that have extra chromosomes or that lack one or more chromosomes. The condition is known as **aneuploidy.** Again there is a change in DNA amount. Sometimes a part of a chromosome breaks off and is lost, or it may become attached to one of the other chromosomes. Then during meiosis one of the daughter cells receives an extra bit of DNA and the other gets less. When the members of the chromosome pairs come together in meiosis, they may exchange parts and the exchange may be unequal—one member of the pair gets more and the other less. There may be serial duplications of segments of DNA along the lengths of the chromosomes, or there may be deletions. It has also been suggested that there may be lateral duplications of the DNA molecule so that the chromosomes contain multiple strands. Thus there are many ways in which changes in the amount of DNA per nucleus can be brought about.

Polyploidy is probably the most successful method of adding new genetic material. The genetic endowment of an organism is made up of a balanced system of interacting parts. Anything that upsets the balance is apt to be deleterious. The best known example of this in humans is Down's syndrome (mongolism), in which the addition of one small extra chromosome results in extreme mental retardation. In polyploidy the balance is not upset because all parts are duplicated.

Increases or decreases in the amount of DNA per nucleus have consequences other than the gain or loss of genetic material. For one thing, there is a good correlation between amount of DNA, the size of the nucleus, and the size of the cell. The more DNA, the larger the nucleus and the larger the cell. Large cells tend to have lower metabolic rates than do small ones, it takes them longer to divide, and the rate of embryonic development is slower when cell size is larger. Polyploid plants are usually larger, with showier flowers and thicker leaves, than diploid ones. The correlation between polyploidy and total size does not hold for animals, in which a polyploid species is usually about the same size as the ancestral one from which it sprang. Apparently increased cell size is compensated by a reduction in the total number of cells.

The various changes in the amount of DNA per nucleus do not seem to happen entirely by chance. They can be caused, or at least their frequency of occurrence can be very much increased, by certain environmental factors. For one thing, radiation can cause chromosome breakage, with subsequent loss of part of the DNA.

It seems probable that sex hormones, or at least the factors that trigger increased production of these hormones, can stimulate DNA synthesis that may not be followed by cell division. The body cells of animals sometimes show different amounts of DNA that correspond to different levels of ploidy. In the liver of a single animal there can be cells that have DNA values corresponding to the diploid ($2n$), tetraploid ($4n$), octoploid ($8n$), or even $16n$ amounts. The cells apparently go through successive cycles of DNA synthesis without continuing on through mitosis. We have found that specimens of a number of species of tree frogs, taken in the breeding ponds in breeding condition, have such heightened ploidy levels in the liver. Specimens taken outside the breeding season may show an occasional cell with the $4n$ DNA amount, a reflection of the normal low rate of cell division that takes place in the liver, but these specimens can be clearly distinguished from the breeding animals (see Table 6-1).

Anything that tends to disrupt the ordinary, and orderly, processes of mitosis can result in a cell with a changed amount of DNA. **Interphase,** the period before the onset of the obvious changes that lead to cell division, can be divided into three stages, G_1, **S,** and G_2. During G_1, much protein synthesis is going on and the cell usually grows in size. In S, the DNA is duplicated. Stage S is followed by G_2. Just what is happening during G_2 is not yet clear, but apparently it is a stage during which a good deal of energy is either used or stored for the succeeding events of mitosis. In **prophase** the chromosomes contract; they become shorter and thicker and can be seen under the microscope as discrete bodies, each composed of two similar **chromatids** joined to each other at the **kinetochore.** A spindle-shaped body made up of microtubules of protein

Table 6–1. Summary of DNA measurements in hylid frogs.

Species	Sex, age	Polyploid nuclei (per cent)			
		2n	4n	8n	16n
Acris gryllus	Early breeding male	87	13	0	0
Hyla cinerea	Adult male	100	0	0	0
Hyla cinerea	Adult female	100	0	0	0
Hyla cinerea	Adult female	96	4	0	0
Hyla crucifer	Breeding female	64	32	4	0
Hyla crucifer	Breeding male	100	0	0	0
Hyla crucifer	Young male	98	2	0	0
Hyla crucifer	Breeding male	100	0	0	0
Hyla crucifer	Breeding male	100	0	0	0
Hyla gratiosa	Young female	100	0	0	0
Hyla ocularis	Adult male	100	0	0	0
Hyla squirella	Adult	100	0	0	0
Pseudacris nigrita	Breeding female	76	24	0	0
Pseudacris nigrita	Young male	96	4	0	0
Pseudacris ornata	Breeding female	44	40	16	0
Pseudacris ornata	Breeding male	60	33	7	0
Pseudacris triseriata	Adult male	98	2	0	0
Pseudacris triseriata	Breeding female	44	29	24	3

takes shape, the nuclear membrane disappears, and the chromosomes move to the equatorial plane of the spindle. At **metaphase** the chromosomes are aligned on the equatorial plate and are attached to the fibers that make up the spindle. The two halves of each chromosome are still joined together. At **anaphase** the kinetochores divide and the chromatids begin to move apart. There are now two clusters of daughter chromosomes, each comprising one set of chromatids. During **telo-**

phase the chromosomes uncoil to form a tangle of long threads, individual chromosomes can no longer be distinguished, nuclear membranes form around each cluster, and the cell divides.

A great many things are known that can interrupt the mitotic cycle at different stages. Various chemicals that act as respiratory poisons can hold the cell in the G_2 stage, after DNA synthesis has been completed but before the onset of mitosis. In the presence of some substances, such as antibiotics, the disintegration of the nuclear membrane during prophase is delayed or does not take place. The chromosomes condense but the kinetochores do not divide. If the nuclear boundary later breaks down, there may be a disorderly scattering of the chromosomes followed by a division of the cell into two daughter cells with unequal amounts of DNA. If the nuclear membrane is retained, the chromosomes return to the spread-out telophase condition. The result is a cell with double the amount of DNA.

Still other chemicals, such as colchicine, can prevent the formation of the spindle or, if it is already formed, can cause it to break down. The kinetochores divide but the chromatids either remain clumped together or separate in a disorderly fashion. Again the result may be the formation of a polyploid cell or of daughter cells with unequal numbers of chromosomes and hence unequal amounts of DNA. Since polyploid plants are often more desirable than their diploid ancestors, with showier flowers and/or larger fruit, horticulturists frequently use cholchicine to induce polyploid formation.

Here, though, we are primarily interested in natural conditions that cause disruptions in mitosis. Marked fluctuations in temperature can cause a disruption. Lack of oxygen and the presence of large amounts of inorganic phosphates can cause poisoning in the prophase which results in the formation of polyploid cells. These three conditions are found in shallow, freshwater ponds and swamps that are rich in vegetation. Temperature fluctuations are much more extreme in shallow water than in deep water. Ponds tend to trap minerals, including phosphate, washed into them from the surrounding land. These provide nutrients needed for plant growth. When the vegetation dies and sinks to the bottom, it is attacked by decay bacteria, which break down the organic remains into inorganic substances. At the same time, the bacteria use up much of the oxygen dissolved in the water. Mud, a mixture of mineral particles, organic remains, and water, tends to accumulate on the bottom of such ponds; the water is often turbid and is rich in nutrients such as phosphate but low in oxygen. The conditions in these ponds might encourage the development of polyploidy in the plants and animals living there.

Another factor should also be considered. As we pointed out, a strong selection pressure acts to eliminate excess DNA. For an organism to make a major adaptive shift, it must not only acquire uncommitted DNA, it must retain it long enough to allow for the slow accumulation of the needed mutations. There should be selective forces in the environment that favor the retention of large amounts of DNA and so counter the forces that favor its loss. It has been suggested that, in any one group, species that have large amounts of DNA and large cells have lower metabolic rates than do those with less DNA and smaller cells. A lower metabolic rate means that less oxygen is consumed during a given period. Thus a habitat in which the oxygen supply is limited may selectively favor organisms with large amounts of DNA. Among ocean-dwelling fishes, those that live at great depths tend to have more DNA than those that live at the surface. Freshwater fishes that live in shallow, vegetation-rich waters include the mud minnow of the eastern United States, which has a DNA value of 5.4

picograms; the goldfish, 4.0; the carp, 3.4; and a South American catfish, 8.8. Compare these with the modal value of 1.7 for the teleosts as a whole. The Congo eel, the salamander with the highest DNA value of any tetrapod, lives in swamps and lowland pools and ditches. These data are not sufficient to prove that there is a correlation between shallow, vegetation-rich, oxygen-poor waters and high DNA content; they only suggest that such a correlation may exist and that it would be worth looking for.

Because of the nature of the sediments in which they are found and because of their structural adaptations, many paleontologists believe that the rhipidistian ancestors of the tetrapods lived in just such waters. We do not know where the ancestors of the land plants and the terrestrial arthropods lived; we do know that today shallow fresh waters support rich growths of algae and aquatic arthropods.

In summary, then, we believe that the ancestors of the inhabitants of the land moved from the oceans into estuaries and rivers, then into shallow lakes and ponds; that they acquired large amounts of uncommitted DNA; and that this endowed them with the genetic versatility to make the journey onto land. This is little more than a guess now, but someday we may know more.

SELECTED REFERENCES

BACHMANN, K., B. A. HARRINGTON, and J. P. CRAIG. Genome Size in Birds. *Chromosoma,* **37,** pp. 405–416, 1972.

SMITH, H. H., ed. *Evolution of Genetic Systems* (Brookhaven Symposium in Biology 23). New York, Gordon and Breach, Science Publishers, Inc., 1972.

SZARSKI, H. Changes in the Amount of DNA in Cell Nuclei

During Vertebrate Evolution. *Nature,* **226,** no. 5246, pp. 651–652, 1970.

WILSON, G. B. *Cell Division and the Mitotic Cycle.* New York, Van Nostrand Reinhold Company, 1966.

Geologic Time

Era	Period	Time from beginning in millions of years	Major events
Cenozoic	Quaternary	2	*Rise and domination of Man*
	Tertiary	63	*Radiation of birds, mammals, and flowering plants*
Mesozoic	Cretaceous	130	*Extinction of most reptile stocks, angiosperms becoming dominant*
	Jurassic	181	*Origin of birds; appearance of angiosperms*
	Triassic	230	*Radiation of reptiles; appearance of mammals; increase in gymnosperms*

	Permian	280	*Amphibians decline as reptiles spread, modern insects appear; conifers spread as primitive land plants decline*
	Pennsylvanian	320	*First reptiles, climax of amphibians, land snails present, extensive coal forests*
	Mississippian	345	*Radiation of amphibians; primitive gymnosperms*
Paleozoic	Devonian	405	*First vascular plants, first forests, first insects and arachnids; amphibians appear at close of period*
	Silurian	425	*First land plants and (?) millipedes*
	Ordovician	500	*Earliest known vertebrates*
	Cambrian	600	*Marine algae and major invertebrate phyla present*
Precambrian		3600±	*Beginning of life; marine algae and some invertebrate animals appear*

INDEX

Italicized numbers refer to figures.